Applied Soft Computing and Embedded System Applications in Solar Energy

Mathematical Engineering, Manufacturing, and Management Sciences

Series Editor:

Mangey Ram

Research Professor, Department of Mathematics; Computer Science & Engineering, Graphic Era Deemed to be University, Dehradun, India.

The aim of this new book series is to publish the research studies and articles that bring up the latest development and research applied to mathematics and its applications in the manufacturing and management sciences areas. Mathematical tool and techniques are the strength of engineering sciences. They form the common foundation of all novel disciplines as engineering evolves and develops. The series will include a comprehensive range of applied mathematics and its application in engineering areas such as optimization techniques, mathematical modelling and simulation, stochastic processes and systems engineering, safety-critical system performance, system safety, system security, high assurance software architecture and design, mathematical modelling in environmental safety sciences, finite element methods, differential equations, reliability engineering, etc.

Partial Differential Equations: An Introduction
Nita H. Shah and Mrudul Y. Jani

Linear Transformation
Examples and Solutions
Nita H. Shah and Urmila B. Chaudhari

Matrix and Determinant
Fundamentals and Applications
Nita H. Shah and Foram A. Thakkar

Non-Linear Programming
A Basic Introduction
Nita H. Shah and Poonam Prakash Mishra

Applied Soft Computing and Embedded System Applications in Solar Energy
Rupendra Kumar Pachauri, J. K. Pandey, Abhishek Sharma, Om Prakash Nautiyal, and Mangey Ram

For more information about this series, please visit: https://www.routledge.com/ Mathematical-Engineering-Manufacturing-and-Management-Sciences/book-series/ CRCMEMMS

Applied Soft Computing and Embedded System Applications in Solar Energy

Edited by
Rupendra Kumar Pachauri,
Jitendra Kumar Pandey, Abhishek Sharma,
Om Prakash Nautiyal, and Mangey Ram

CRC Press
Taylor & Francis Group
Boca Raton London New York

CRC Press is an imprint of the
Taylor & Francis Group, an **Informa** business

First edition published 2021
by CRC Press
6000 Broken Sound Parkway NW, Suite 300, Boca Raton, FL 33487-2742

and by CRC Press
2 Park Square, Milton Park, Abingdon, Oxon, OX14 4RN

Library of Congress Cataloging-in-Publication Data
Names: Pachauri, Rupendra Kumar, editor. | Pandey, J. K. (Nanotechnology specialist), editor. | Sharma, Abhishek (Robotics engineer), editor. | Nautiyal, Om Prakash, editor. | Ram, Mangey, editor.
Title: Applied soft computing and embedded system applications in solar energy / edited by Rupendra Kumar Pachauri, J. K. Pandey, Abhishek Sharma, Om Prakash Nautiyal, Mangey Ram.
Description: First edition. | Boca Raton, FL : CRC Press, 2021. | Series: Mathematical engineering, manufacturing, and management sciences | Includes bibliographical references and index. | Summary: "This book deals with energy systems and soft-computing methods from a wide range of approaches and application perspectives. The authors examine how embedded system applications can deal with the smart monitoring and controlling of stand-alone and grid connected solar PV systems for increased efficiency"— Provided by publisher.
Identifiers: LCCN 2020054777 (print) | LCCN 2020054778 (ebook) | ISBN 9780367625122 (hardback) | ISBN 9781003121237 (ebook)
Subjects: LCSH: Photovoltaic power systems—Automatic control. | Embedded computer systems—Industrial applications. | Soft computing.
Classification: LCC TK1087 .A73 2021 (print) | LCC TK1087 (ebook) | DDC 621.31/244028563—dc23
LC record available at https://lccn.loc.gov/2020054777
LC ebook record available at https://lccn.loc.gov/2020054778

ISBN: 978-0-367-62512-2 (hbk)
ISBN: 978-0-367-63902-0 (pbk)
ISBN: 978-1-003-12123-7 (ebk)

Typeset in Times
by codeMantra

Contents

Preface

This book is motivated by the desire that we and others must pursue the evolution of the core course of solar energy. Many institutions around the world have revised their curriculum in response to the introductory course on solar energy. This book comprises the two themes of soft computing and embedded system applications in solar energy. The first objective of this proposed book is to provide an initial knowledge of solar energy, which provides a solid foundation for further study. However, soft computing technology is becoming increasingly important in a much wider range of scientific and engineering disciplines. The second objective, therefore, is to give those students who do not take advanced courses in the embedded system conceptual tools the necessary information in this field. Finally, the broader objective is to expose all students not only to programming concepts but also to the experimentally rich foundations of the energy field.

MATLAB® is a registered trademark of The MathWorks, Inc. For product information, please contact:

The MathWorks, Inc.
3 Apple Hill Drive
Natick, MA 01760-2098 USA
Tel: 508-647-7000
Fax: 508-647-7001
E-mail: info@mathworks.com
Web: www.mathworks.com

Acknowledgements

The editors of this book are very grateful to many students and colleagues who have provided extensive feedback and advice in developing this text. It evolved over many years through the development and teaching of courses at University of Petroleum & Energy Studies, Graphic Era University, and Uttarakhand Science Education and Research Centre. These universities have been very supportive throughout the preparation of this book.

Many ideas and explanations throughout the book were inspired through numerous collaborations. For this reason, the editors are particularly grateful to the helpful insights and discussions that arose through corroboration with international researchers in the specified area. Over the years of interaction, their ideas helped the editors to shape their perspective and change their presentation throughout the book.

Finally, the editors are very thankful to the people involved with CRC Press/ Taylor & Francis Group for their efforts and advice in preparing the manuscript for publication.

Editors

Dr. Rupendra Kumar Pachauri completed his M. Tech from the Electrical Engineering Department, ZHCET, Aligarh Muslim University (AMU), Aligarh, India, in 2009. He received his Ph.D. in renewable energy from G. B. University, Gautam Buddha Nagar, India, in 2016. Presently, he is working as an Assistant Professor – Selection Grade in the Department of Electrical and Electronics Engineering, University of Petroleum and Energy Studies (UPES), Dehradun, India. He has published more than 75 research publications in internationally reputed science citation/ Scopus-indexed journals along with presentations at IEEE/Springer conferences. His fields of research are solar energy, fuel cell technology, applications of ICT in PV system, and smart grid operations.

Prof. (Dr.) Jitendra Kumar Pandey received his Ph.D. in nanotechnology. Presently, he is working as Professor & Head of R&D in the University of Petroleum and Energy Studies (UPES), Dehradun, India and providing strategic support to the University for enhancing the research and development in all five schools (Engineering, Design, Computer, Business, and Law) while working closely with the senior university administration including the Vice Chancellor, Dean Academic, and Deans of the other schools of the University. He received the JSPS fellowship, and his areas of specialization are nanotechnology, 3D printing, and optimization.

Mr. Abhishek Sharma received his Bachelor's degree in electronics and communication engineering from ITM-Gwalior, India, in 2012 and his Master's degree in robotics engineering from the University of Petroleum and Energy Studies (UPES), Dehradun, India, in 2014. He was a Senior Research Fellow in a DST-funded project under the Technology Systems Development Scheme and worked as an Assistant Professor with the Department of Electronics and Instrumentation, UPES (2015–2016). Currently, he is working as a research scientist in research and development department and is also the head of the innovation lab, UPES. He was a research fellow for two years and worked on motion planning and controlling of a swarm of drones in Ariel university, Israel. His research interests include Swarm Robotics, Multi-Objective Optimization, and Swarm Intelligence.

Dr. Om Prakash Nautiyal received his Ph.D. in physics from H.N.B. Garhwal Central University Srinagar, Garhwal, Uttarakhand, India. He also acquired his Master's degree in three different subjects: physics, information technology, and education with first division. He has been a faculty member for around ten years in various capacities and has more than nine years of experience as scientist in Uttarakhand Science Education & Research Centre (USERC), Department of Information and Science Technology, Govt. of Uttarakhand. Given his approach to leveraging science and technology, he has always been involved in developing scientific and technological solutions and innovations that are beneficial to the society at large and particularly to the last-mile learners in difficult geographies like Uttarakhand in India. Dr. Om Prakash is on the editorial broad of international and national journals of repute. He is a regular reviewer for several national and international journals. He has been a member of the organizing committee of a number of international and national conferences, seminars, and workshops. He is working as Principal Investigator and Coordinator of some projects externally funded by the Govt. of India, BRNS, DAE, DST, and MoES. He has authored/edited 5 books and has more than 30 publications in reputed national and international journals as well book chapters and conference proceedings.

Prof. (Dr.) Mangey Ram received his Ph.D. in mathematics, with a minor in computer science from G. B. Pant University of Agriculture and Technology, Pantnagar, India. He has been a Faculty Member for around twelve years and has taught several core courses in pure and applied mathematics at undergraduate, postgraduate, and doctorate levels. He is currently the Research Professor at Graphic Era (Deemed to be University), Dehradun, India. Before joining the Graphic Era, he was a Deputy Manager (Probationary Officer) with Syndicate Bank for a short period. He is Editor-in-Chief of the *International Journal of Mathematical, Engineering and Management Sciences, Journal of Reliability and Statistical Studies* and Editor-in-Chief of six book series with Elsevier, CRC Press/Taylor & Francis Group, Walter De Gruyter Publisher, Germany, River Publisher, and the Guest Editor and Member of the editorial board of various journals. He has published 225 plus research publications (journal articles/books/book chapters/conference articles) in IEEE, Taylor & Francis, Springer, Elsevier, Emerald, World Scientific, and many other national and international journals and has also presented at conferences. Also, he has published more than 50 books (authored/edited) with international publishers like Elsevier, Springer Nature, CRC Press/Taylor & Francis Group, Walter De Gruyter Publisher Germany, and River Publisher. His fields of research are reliability theory and applied mathematics. Dr. Ram is a Senior Member of the IEEE, Senior Life Member of Operational Research Society of India, Society for Reliability Engineering, Quality and Operations Management in India, and Indian Society of Industrial and Applied Mathematics. He has been a member of the organizing committee of a number of international and national conferences, seminars, and workshops. He has been conferred with "Young Scientist Award" by the Uttarakhand State Council for Science and Technology, Dehradun, in 2009. He has been awarded the "Best Faculty Award" in 2011; "Research Excellence Award" in 2015; and recently "Outstanding Researcher Award" in 2018 for his significant contribution to academics and research at Graphic Era Deemed to be University, Dehradun, India.

Contributors

Moshe Averbukh
Department of Electrical
    and Electronics
Ariel University
Ariel, Israel

M. Gurunadha Babu
Department of Electronics and
    Communication
Holy Mary Institute of Technology
    and Science
Hyderabad, India

Manish Bilgaye
Department of Electronics
    and Communication
Holy Mary Institute of Technology
    and Science
Hyderabad, India

Yogesh K. Chauhan
Department of Electrical Engineering
Kamla Nehru Institute of Technology
Sultanpur, India

Ankit Dasgotra
Department of Research
    & Development
University of Petroleum & Energy
    Studies
Dehradun, India

Santosh Ghosh
Corporate R&D Department
Kirloskar Brothers Limited
Pune, India

Anitya Kumar Gupta
Masters Computer Science
    Data Analytics
Illinois Institute of Technology
Chicago, Illinois

Ankur Kumar Gupta
Research and Development Department
    (SOCSA)
IIMT University
Meerut, India

Peeyush Kala
Department of Electrical Engineering
Women's Institute of Technology
Dehradun, India

Safia A. Kazmi
Department of Electrical Engineering
ZHCET, Aligarh Muslim University
Aligarh, India

Adesh Kumar
Department of Electrical and
    Electronics
University of Petroleum and Energy
    Studies
Dehradun, India

R. Vijaya Kumar
Department of Ocean Engineering
Indian Institute of Technology IIT(M)
Chennai, India

Raj Kumar Mishra
Department of Mechanical & Industrial
    Engineering
IIT Roorkee
Roorkee, India

R. L. Meena
Department of Electrical Engineering
Delhi Technological University (DTU)
Delhi, India

R. Raajiv Menon
Department of R&D
University of Petroleum and Energy
    Studies
Dehradun, India

Reetu Naudiyal
Department of Electrical Engineering
University Polytechnic, Uttaranchal
    University
Dehradun, India

Rupendra Kumar Pachauri
Electrical and Electronics Engineering
SOE University of Petroleum and
    Energy Studies
Dehradun, India

Jitendra Kumar Pandey
Department of R&D
University of Petroleum and Energy
    Studies
Dehradun, India

Vikas Pandey
Electrical Engineering Department
Babu Banarasi Das University
Lucknow, India

Manjaree Pandit
Electrical Department
Madhav Institute of Technology
    & Science
Gwalior, India

Karni Pratap Palawat
Department of Electrical Engineering
Delhi Technological University (DTU)
Delhi, India

Shailendra Rajput
Department of Electrical
    and Electronics
Ariel University
Ariel, Israel

Y. David Solomon Raju
Department of Electronics and
    Communication
Holy Mary Institute of Technology
    and Science
Hyderabad, India

Sandeep Rawat
Department of Electrical Engineering
University Polytechnic, Uttaranchal
    University
Dehradun, India

Devender Saini
Electrical & Electronics Department
University of Petroleum & Energy
    Studies
Dehradun, India

Akhilesh Sharma
Electrical Engineering Department
NERIST
Nirjuli, India

Aayush Shrivastava
Electrical & Electronics Department
University of Petroleum & Energy
    Studies
Dehradun, India

Arun Kumar Singh
Saudi Electronic University
Saudi Arabia-KSA

Vishal Kumar Singh
Department of Research
  & Development
University of Petroleum & Energy
  Studies
Dehradun, India

Sunil Kumar Tiwari
Department of Research &
  Development
University of Petroleum & Energy
  Studies
Dehradun, India

Nafees Uddin
Department of Applied Science
JIMS Engineering Management
  Technical Campus
Greater Noida, India

Vinod Kumar Yadav
Department of Electrical
  Engineering
Delhi Technological University (DTU)
Delhi, India

1 MPPT Control Systems for PV Power Plants

Shailendra Rajput and Moshe Averbukh
Ariel University, Ariel, Israel

CONTENTS

1.1 INTRODUCTION

1.1.1 ECONOMIC ASPECTS OF PV POWER SOURCES

Solar energy production has increased substantially during the past decade (Figure 1.1). This significant growth is because this is one of the cheapest sources of electricity. Graphical representations of price declination for 1 MWh of power from PV models, compared with other energy sources, can explain this situation. In 2014–2015, PV electricity cost achieved grid-parity; nowadays, they are 1.2–2.5 times cheaper than electricity from other types of energy plants. Owing to this tendency of reduced PV prices, the solar electricity market is a fast-growing trading sector in the global economy and is, today, estimated at more than $10 billion/year, with the increasing potential of further increase. Most economic prognoses predict further continuation in price reduction that makes PV solar electricity the most outstanding and the most promising of power sources in the near future (Figure 1.2).

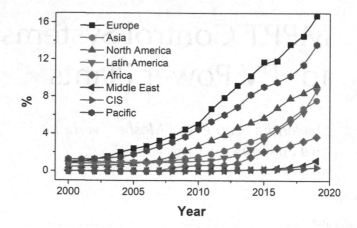

FIGURE 1.1   Relative growth of solar electricity production during the past decade [1].

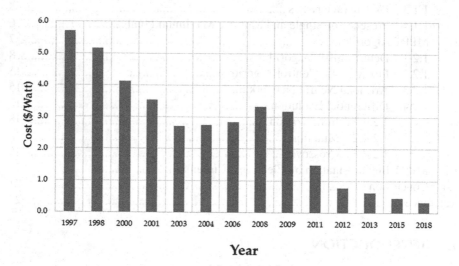

FIGURE 1.2   The cost reduction of PV modules per watt of power [2].

1.1.2   Major Requirements for PV Facilities Control

The most common request for facilities control can be expressed by the economic reason assuming a minimum of Levelized Cost of Electricity (LCOE) [3]:

$$\text{LCOE} = \frac{\text{Total Life Cycle Cost}}{\text{Total Lifetime Energy Production}} \qquad (1.1)$$

The numerator of the expression (1.1) represents some constant and negligibly variable parameters. The denominator represents, the total lifetime of PV modules, which is accounted for today as ~25 years of service life. The service life of PV modules depends not only on the semiconductor manufacturing technology, but environmental

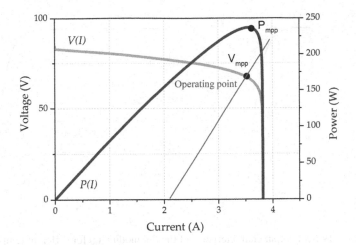

FIGURE 1.3 Typical I-V and P-I curves of the PV module.

conditions also affect the service life of PV modules. Consequently, total life costs (including the initial cost, transportation, installation, and maintenance) may be assumed as a constant value. Therefore, the LCOE is mainly influenced by the total lifetime energy, which should be maximized to achieve minimum LCOE.

Output characteristics of both the PV module and the load are necessary to determine the real electricity production. It is well known that PV cells generate DC current. Hence, a special DC–AC inverter should be applied to use the AC load in modern networks. Taken together, all of this leads to the need to estimate the PV plant's functionality connected to variable loads. An analytical description of the output characteristics of PV modules is required.

1.1.3 PV Characteristics

Typical output characteristics of the PV module are presented in Figure 1.4. It can be seen that the output power depends on solar irradiation and temperature. PV systems follow the law of energy conservation. Power is a derivative of the solar energy that is converted to electrical energy using the principles of semiconductors.

The variable $I–V$ curve points out that power generation from the PV modules does not remain constant. PV power depends not only on the solar intensity and temperature but also on the characteristics of load. The operating point representing the voltage–current combination is at the intersection of the panel, and load characteristics determine output, and the power should be maximized. The operating or working point can be found graphically or analytically to solve the equation of output characteristics $(I = f(V))$ for the PV module and for a given load. The analytical solution requires strict description of a function $I = f(V)$ for a module. The equivalent PV circuit can utilized for this purpose. There are different representations of the equivalent circuit [5]. Two types of equivalent circuits are mostly used: single-diode and double-diode models (Figure 1.5a and b).

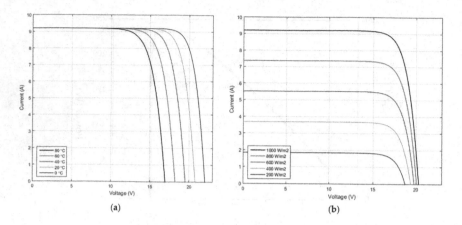

FIGURE 1.4 Typical output characteristics of the PV module (a) for different temperatures, (b) for different solar irradiations.

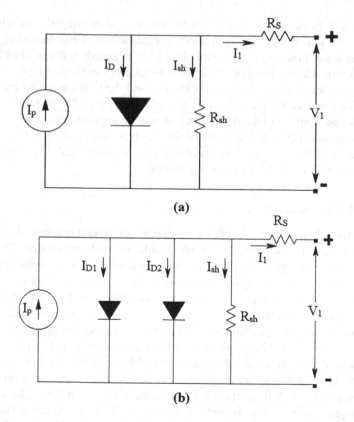

FIGURE 1.5 (a) Single-diode equivalent circuit, and (b) double-diode equivalent circuit (I_p: photocurrent proportion to solar intensity and temperature, I_D: diode current, R_s, R_{sh}: serial and shunt equivalent resistances).

The equivalent circuit is analyzed using the Kirchhoff circuit law. For a single-diode circuit,

$$I_l = I_p - I_{D_1} - I_{sh} \tag{1.2}$$

where I_l is the load current, I_{D_1} is the diode current, and I_{sh} is the current of the shunt resistance. The diode current is expressed using the Shockley equation:

$$I_{D_1} = I_{SD}\left[\exp\left(\frac{q(V_l + I_l R_s)}{a_1 k_B T}\right) - 1\right] \tag{1.3}$$

where I_{SD}, V_l, a_1, k_B, T and q are the reverse saturation current, cell output voltage, diode constant, Boltzmann constant $(1.3806503 \times 10^{23}\,\text{J/K})$, junction temperature (K), and the electron charge $(1.60217646 \times 10^{-19}\,\text{C})$, respectively. The current in shunt resistance is given by:

$$I_{sh} = \frac{V_l + I_l R_s}{R_{sh}} \tag{1.4}$$

After combining equations (1.2)–(1.4), we get:

$$I_l = I_p - I_{SD}\left[\exp\left(\frac{q(V_l + I_l R_s)}{a_1 k_B T}\right) - 1\right] - \frac{V_l + I_l R_s}{R_{sh}} \tag{1.5}$$

The expression (1.5) represents the output characteristic of the PV module. For double-diode circuit, Equations (1.2) can be written as:

$$I_l = I_p - I_{D_1} - I_{D_2} - I_{sh} \tag{1.6}$$

Further,

$$I_l = I_p - I_{SD_1}\left[\exp\left(\frac{q(V_l + I_l R_s)}{a_1 k_B T}\right) - 1\right] - I_{SD_2}\left[\exp\left(\frac{q(V_l + I_l R_s)}{a_2 k_B T}\right) - 1\right] - \frac{V_l + I_l R_s}{R_{sh}}$$

$$\tag{1.7}$$

where I_{SD_1} and I_{SD_2} represent the diffusion and saturation current, respectively. I_{D_1} and I_{D_2} represent the first and double diode current. I_l and V_l characterize the module output as variable parameters. Other parameters must be determined and assessed before the application of Equations (1.5) and (1.7). There are plenty of literature resources dedicated to this purpose [6–8]. The operating point depends on the parameters of the load and the current characteristic of the PV module and can be positioned in each place on the I–V curve. There is a special point (V_{mpp}, I_{mpp}) at which the output power becomes maximum (P_{mpp}). The optimal parameters (V_{mpp}, I_{mpp}) constantly change depending on the intensity of the solar irradiation, the temperature changes through the day, and the variation in intensity in different seasons in a year. The PV facilities should be connected with an electronic control circuit to generate maximum energy. For example, the DC–DC converter matches the load to the PV module and ensures maximum power generation from PV modules. In other words, the operating point changes corresponding to the variability of the MPP location. These electronic devices ensure maximum power in each moment and is hence called the maximum power point tracker (MPPT). All these electronic devices are designed using the DC–DC converters. However, they are based on different functional algorithms and different circuit topologies.

1.1.4 PRESENCE OF SINGLE AND MULTIPLE MAXIMUM POINTS

The typical PV module has a restricted outcome with a maximum voltage of ~40 V and a current of ~8–10 A. Several modules are connected serially in an array to provide significantly more power. Further, such arrays may be oriented or arranged in parallel strings connected to an MPPT. The array characteristics of an n-serially connected PV module array may differ significantly from those of the individual module. If the PV array comprises evenly irradiated individual modules, the output characteristic has one maximum only (Figure 1.4). However, a large array with tens of PV modules may be irradiated unevenly. Such events take place during cloudy weather when some of the modules can be shadowed. Those modules that received lower dose of irradiation prevent nominal current flow leading to diminished photocurrent (Figure 1.5). Therefore, the simple design of the n-serially connected PV module array is not applicable because the output power drops drastically for uneven irradiations. To prevent such shortcomings, each module is equipped with a bypass diode to avoid any disadvantages arising due to partial shading (Figure 1.6). If some modules are illuminated weakly, then the photocurrent $\left(I_{ph}\right)$ decreases significantly.

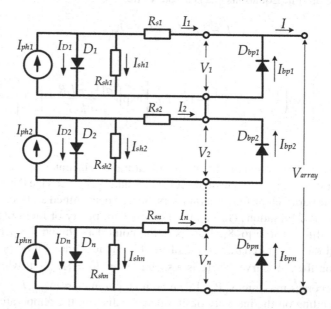

FIGURE 1.6 n-Serially connected PV modules represented by the equivalent circuit and equipped with bypass diodes $\left(D_{bp}\right)$.

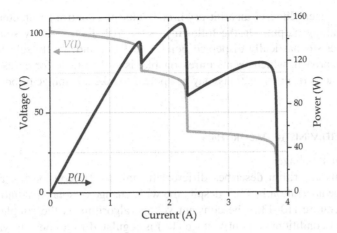

FIGURE 1.7 The presence of local maxima in the characteristic of the unevenly irradiated solar array.

However, the array continues to supply nominal current to bypass the weak module through the bypass diode. This solution allows for the continuation of the electrical power from normally irradiated modules. As a result, the output voltage and characteristics of the array are changed so that instead of a single maximum, several maxima (two, three, or more) can appear in the output characteristic (Figure 1.7). A detailed explanation for the presence of multiple maxima is obtained by considering the simplified equivalent circuits of PV modules [9]. It should be emphasized that single or plural maxima for the MPPT algorithms is important. The control system must find the global maximum (GM) and then keep operating on it.

1.2 MPPT ALGORITHMS

It is well known that PV power generation strongly depends on solar irradiation and module temperature. Hence, MPPT algorithms should permanently search the MPP position and operate on it. The MPPT algorithms can be divided into two main groups based on deterministic and stochastic ways of functionality. The first type operates by the sequential, step-by-step change of a control parameter leading to the optimal point. Considering the functional relation between current and voltage is determined by the module output characteristic, as a control variable that may be chosen is only current or voltage. Therefore, the detailed electronic scheme should be designed, considering whether the parameter of current or voltage was selected for control. Sometimes, the electronic circuit realizes control by regulating some subsidiary parameters (pulse-width modulation, for example), indirectly determining the desirable outcome – current, voltage, or power. There may be two types of MPPT algorithms: deterministic and stochastic. The deterministic approach assumes strict rules for finding the maximum. Algorithms with purely deterministic approaches are applicable only for the single-maxima case.

On the contrary, for the first algorithm type, the stochastic approach of the MPPT functionality is based on the probabilistic selection of a control parameter.

The MPPT system chooses an arbitrary initial value of a control parameter and verifies the resulting output. On the following steps, MPPT chooses by some special rules another stochastically dispersed point, checks it, and finally selects the best one. This approach, which seems unreasonable, is still suitable for cases of partial shading. Synergetic combinations may be applied to take advantage of both types of algorithms.

1.2.1 DETERMINISTIC ALGORITHMS

a. Constant voltage

This algorithm describes different techniques by considering a small voltage aberration in MPP despite the wide change of solar irradiation and temperature [10–13]. The constant voltage algorithm is the simplest one. For all conditions, output voltage (V_O) is regulated to a constant value as a constant ratio to the measured open-circuit voltage $(V_O = K \cdot V_{OC})$. The fixed ratio of coefficient K (~0.7–0.8) as a rule is used. The difficulties in permanently measuring V_{OC} lead to a simplified technique providing constant output voltage permanently. This method can provide significant output power, but the average power is lower than the maximum one. This approach can be improved, and the average energy will be larger when temperature changes are considered. It is assumed that the change of controlled V_O by the mathematical expression approximating the change of a desired voltage magnitude is given as:

$$V_{mpp}(T) = V_{mpp}(T_{ref}) + K_{U_{mpp}}(T - T_{ref}) \qquad (1.8)$$

where $V_{mpp}(T)$ is the desired value of V_O under the present temperature T of a module, $V_{mpp}(T_{ref})$ is V_{mpp} under nominal temperature T_{ref}, and $K_{U_{mpp}}$ is the temperature coefficient of V_{mpp}. Similar to constant voltage, the constant current algorithm was also studied [14]. The same idea to relate I_{mpp} with I_{SC} by the special coefficient is also considered. However, I_{SC} alters much more significantly with the variation of solar irradiation intensity compared to the change of V_{OC}. Therefore, constant current control is less effective than that of constant voltage control. Moreover, it is inapplicable for low illuminations since these conditions can lead to a drastic decrease of PV output.

b. Perturb and Observe algorithms

Perturb and Observe (P&O) algorithms are the most used MPPT algorithms [15–18]. This method is based on a simple feedback arrangement and corresponding measurements of the output parameters. As for this method, the module voltage or current (one of which is a control variable) is periodically perturbed a little. The corresponding output or its derivative is compared with the previously obtained outcome, and the following perturbation is chosen in accordance with the previous results. The flowchart for this method is shown in Figure 1.8. The important issue with this approach is the stability of the control process. To ensure this, relatively slight perturbations are applied that, on the one hand, increase the stability

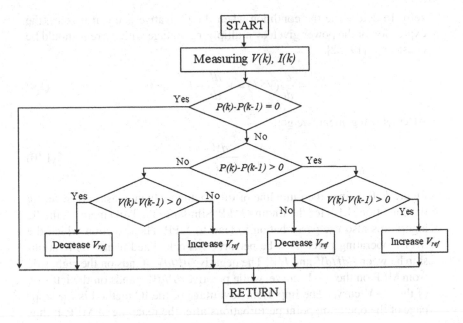

FIGURE 1.8 Flowchart of P&O method.

but, on the other hand, cause elongation of the MPPT detection. In case the power is increased during the previous step or the derivative is positive, the perturbation continues in the same direction. Otherwise, when the derivative becomes negative or the power is diminished, the perturbation changes direction. After the peak power is reached, the power change in the vicinity of MPP obtains an oscillating behavior since every perturbation causes a negative change of power. Now when the peak point is reached, the power change in the vicinity of MPP obtains oscillating behavior since every perturbation causes a negative change in power. The schematic demonstration of the P&O approach is shown in Figure 1.9.

c. Incremental Conductance

The incremental conductance (IC) method is based on the fact that the derivative of the change in the power at the point of a maximum is equal to

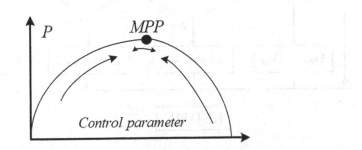

FIGURE 1.9 Dynamic process of MPP achievement as per the P&O method.

zero. To determine the condition when the derivative is equal to zero, the expression of the power given by multiplying voltage with current should be considered [19–22].

$$\frac{dP}{dt} = \frac{d}{dt}(V \cdot I) = V\frac{dI}{dt} + I\frac{dV}{dt} = 0 \qquad (1.9)$$

After rearrangement, we get:

$$\frac{V}{I} = -\frac{dV}{dI} \qquad (1.10)$$

Figure 1.10 defines the guideline of this method which consists of meeting the condition (1.10) for the finding MPP. Similar to the P&O method, the IC algorithms also use perturbation to lead to MPP. The direction which the MPPT operating point must be perturbed is calculated using the relationship between $-dI/dV$ and I/V. The negative dP/dV stands on the right side from MPP on the $P - V$ curve, while positive dP/dV stands on the left side of the $P - V$ curve. The important advantage of the IC method is the stoppage of the operating point perturbations after the reaching of MPP. It thus

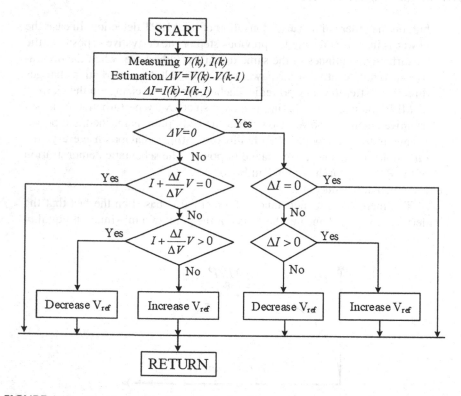

FIGURE 1.10   Flowchart of the IC method algorithm.

offers a significant benefit compared with P&O. The disadvantage of this algorithm is its increased complexity.

d. Predictive control

None of the above-mentioned algorithms can be efficiently applied in partial shading conditions. During partial shading, the V–I characteristic is in the form of a complex step-like configuration, causing the presence of multiple local MPP on the P–I curve. If the first MPP is the global one, then the algorithm will be effective. However, if the first local maximum may not be a global one, the control system following this approach consists of finding and then keeping this first maximum that will then result in lower average output. Contrary to previous approaches, the main idea of predictive control is to estimate the position of GMPP and lead the control parameter to the vicinity of the optimal value, bypassing local maxima, and further to apply P&O or the IC method to determine global maximum with high precision. The combination of predictive control with some of the deterministic approaches based on a sequential search for MPP can be significantly more efficient than that of using any one separately. In this way, predictive control can be much more accurate, serving only to point out the range of the GMPP location [23]. The proposed method for efficiency uses a simplified equivalent circuit of solar modules (Figure 1.11a). The output characteristics of an array are shown in Figures 1.11b–d. Figure 1.11a represents the simplified equivalent circuit, including the source photocurrent (I_{ph}) and the shunt diode (D). Shunt resistance (R_{sh}) is assumed to be infinite, and serial resistance (R_s) is equal to zero. The V–I and P–I curves corresponding to this equivalent circuit are shown in Figure 1.11b. Figure 1.11c and d shows the output characteristics of the n-serially connected PV module array. The array comprises unevenly irradiated PV modules. Hence, multiple power maxima exist in the P–I curve.

Determination of a simplified equivalent circuit requires permanent measurements of voltage and current magnitudes of each module.

e. Ripple correlation current

The algorithm consumes the artificially created ripple of a control variable (e.g. current) with the following derivatives: comparison of power and the control variable [24]. The control system design based on this approach is simplified due to the DC–DC converter's inevitable use in any MPPT system. The main principle of the DC–DC converter is the switching of the current flow. Therefore, a ripple of a control parameter is the inherent property and is seen automatically during MPPT functionality. The description of this method of operation is as follows.

If the operating point is on the left side of MPP on the P–I curve, then the current/power increments and decrements that are produced by a ripple are coinciding in sign. However, they are opposite in sign on the right side of the MPP. They are coinciding in sign for half-period of a ripple and are opposite in the other half-period at the MPP. Therefore, the multiplication

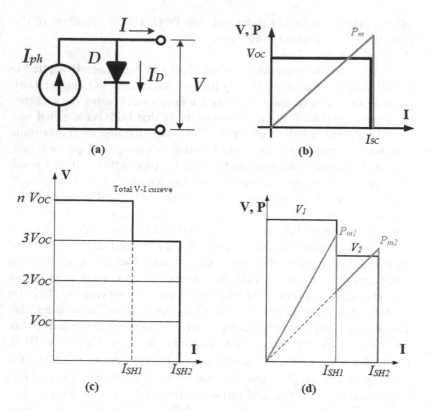

FIGURE 1.11 (a) Simplified equivalent circuit, (b) *V–I* and *P–I* curves of a simplified circuit, (c) *V–I* curve of *n*-serially connected identical PV modules when one or more of them are illuminated weaker, and (d) *P–I* curve having several maxima of *n*-serially connected unevenly illuminated modules.

of the derivative of a current (*di/dt*) and derivative of power (*dP/dt*) can provide the current control direction. The control for the current increment can be formulated as follows:

$$\Delta i = K_i \cdot \int_0^T \frac{dP}{dt} \cdot \frac{di}{dt} \cdot dt \qquad (1.11)$$

where K_i is the coefficient to determine the rate of current change. The magnitude of K_i agrees with the frequency of the ripple and should be decreased with the increase in ripple frequency. It is worth recalling that the ripple frequency cannot be large. Otherwise, the accurate assessment of the power derivative becomes difficult. If module voltage is used as a control variable, then the same expression (1.11) can be applied with the coefficient K_v. The algorithm behind this principle suffers from a relatively prolonged search time because the ripple frequency should be low in real-world situations. Therefore, such algorithms are not suitable for fast-changing conditions.

Also, electronic implementation of the control is rather complicated and requires a fast and precise estimation of power and its derivative.

f. Current sweep method

The method uses the fast periodic change of current from zero to I_{SC} [25]. During the increase in current, the corresponding voltage is being measured and instantaneous power is being calculated. The derivative of power is assessed during the current sweep, and the optimal current is calculated. Further, the control system regulates the current magnitude at the optimal level. The sweep process should be carried out periodically by considering environmental conditions (the changes in solar power and temperature). This application of this method is restricted to fast-changing conditions. However, it can be very effective to find GMPP during partial shading conditions.

g. Stochastic algorithms

In stochastic algorithms, the process of GMPP search is carried out stochastically and unpredictably. Their functionality is similar to the behavior of living organisms in nature or human mental activity. The major advantage of these algorithms is the ability to adapt, which is especially important in variable reality. Solar modules work for 25–27 years, and their parameters change with aging. Environmental conditions are also changing. Dust and dirt cover the module surfaces gradually and unevenly. Therefore, stochastic algorithms may be superior compared to deterministic approaches in the long term. The following major types of stochastic methods can be found in the literature:

- Fuzzy logic controller
- Particle swarm optimization
- Genetic algorithm
- Differential evolution (DE)
- Artificial neural network

1.2.2 Fuzzy Logic Control Algorithm

This concept applies expert knowledge only and does not need any exact model of control system design to work [26–29]. It provides the possibility of working under varying conditions (solar irradiation, temperature, dust, shading) and considers the aging of the PV equipment. This method applies some linguistic forms such as "many," "low," "medium," "often," and "few" [26]. The control process includes four stages: fuzzification, rule base, inference engine, and defuzzification (Figure 1.12).

The values of the membership function are assigned during the fuzzification stage to linguistic variables. The fuzzy-control system designer determines the number of linguistic variables, which is determined based on the accuracy of the desired output. For example, linguistic variables for the five fuzzy rule bases can be Negative Big (NB), Negative Small (NS), Zero (Z), Positive Small (PS), and Positive Big (PB) [26]. Larger numbers of linguistic variables generate more accurate and stable output characteristics in the design [30], though longer time is needed to get more accurate results. As a rule, the triangular shape is mostly used. However, various

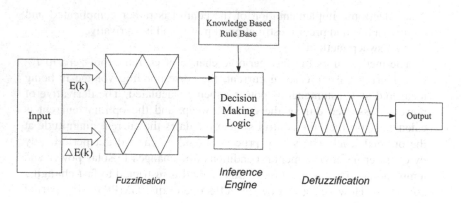

FIGURE 1.12 Four stages of the fuzzy-logic control process [14].

membership function shapes exist such as Gaussian, trapezoidal, and triangular. The range of membership functions, the number of membership functions, and parameters (*a* and *b*) can be decided by known parameters of the proposed scheme [26]. If parameters are unknowns or unachievable, the trial-and-error method, based on user knowledge, can be applied [27]. Of course, this should be such that the input data covers the entire region of interest. The diagrams show that fuzzy inputs are variable, usually having an error (*E*) and change in error (Δ*E*). The procedure to calculate *E* and Δ*E* is flexible and determined by the specific user. However, the approximation for *dP/dV* at MPP can be defined as:

$$E(t) = \frac{P(t) - P(t-1)}{V(t) - V(t-1)} \tag{1.12}$$

$$\Delta E(k) = E(k) - E(k-1) \tag{1.13}$$

where *t* is the sampling time, *P(t)* is the instantaneous power of the PV system, and *V(t)* provides the corresponding instantaneous voltage. The *E(t)* value shows the location of the operating load power point, which may be on the left or right side of the MPP or at the point of the MPP. Δ*E(x)* shows the direction of the operating power point change.

After the inputs of *E* and Δ*E* have been estimated and converted to linguistic variables, the output signal *D* can be generated from the lookup rule base table (Table 1.1).

An additional advantage of FLC is the possibility to adjust the gaps between operating points, aiming at achieving MPPT. For example, the membership function (Figure 1.13) can be created such that it is denser in the middle of the range to get a more accurate output signal [14]. It should be noted that a fuzzy output signal is a consequence of a fuzzy understanding of the behavior of the entire system. It should be noted that the prior user knowledge for the formation of fuzzy rules can be formulated using the IF-THEN principle [29]. The commonly used Mamdani's inference engine method uses the min–max operation in fuzzy operation [30]. Acceleration and deceleration represent environmental changes that affect the system. At the

TABLE 1.1
Rule Base Table for the Fuzzy Logic
Controller [14]

			E		
ΔE	NB	NS	Z	PS	PB
NB	ZE	ZE	NB	NB	NB
NS	ZE	ZE	NS	NS	NS
Z	NS	ZE	ZE	ZE	PS
PS	PS	PS	PS	ZE	ZE
PB	PB	PB	PB	ZE	ZE

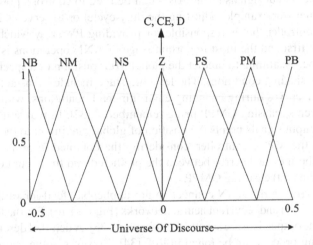

FIGURE 1.13 The membership function of the fuzzy logic controller [14].

defuzzification stage, the FLC output controller converts the linguistic variables to numerical variables using the same membership function but with different user-defined ranges. The FLC output delivers a signal to control the power converter's duty-cycle for the maximum output tracking. Owing to the defuzzification principle, the FLC can maintain the operating voltage (or current) of the PV array at its optimal value, ensuring GMPP even under partial shading conditions. Therefore, the energy output of a PV facility and its efficiency can be improved.

1.2.3 ARTIFICIAL NEURAL NETWORK

The method of control by artificial neural networks (ANN) has significant advantages for solving nonlinear tasks. This leads to its increased use in various fields. It should be noted that the ANN is a system that has rather different control principles than other systems. Instead of having a typical and rigorously established set of deciding rules, it is characterized by a flexible self-organized structure that receives an input, processes the data, and provides output signals. The ANN functions based on sequential learning and need not be reprogrammed when environmental conditions change. It consists of input layers, hidden layers, and output layers. ANN is modeled as weights that connect the neural networks. Figure 1.14 represents the interconnection between i and j assigned by w_{ij} [31]. In other words, an artificial neural network can be seen as an interconnected group of nodes, like neurons in the brain. Each node (designated as a circle) represents an artificial neuron that is connected to another one. The connections between the output of the previous neuron and the input of the following one are represented by arrows.

The structure of ANN is compounded by nodes and interconnections, and is determined by a specific designer. The input variables in the ANN MPPT system can be selected from some PV array characteristics (open-circuit voltage, short-circuit current) and some input arguments such as solar irradiation, temperature, wind velocity, and so on. The output signals of the ANN can be used to control a power converter in some manner (for example, adjusting the duty-cycle) or to serve as input signals for another controller that is responsible for providing PV array which operates at the MPP. The first and the most important stage of ANN operations is the training (learning) process, aiming to predict the global MPP position of a specific PV array under various shading conditions. The learning stage includes constant observation of $P–V$ or $P–I$ curves during changing environmental conditions, with the optimal voltages (currents) causing GMPP being remembered [32]. In real situations, ANN based on the input signals points the position of global maximum to the MPPT controller. Then, the MPPT controller, considering the obtained information, leads to reduction in the inevitable error between the predicted and the accurate parameters of a voltage (current) ensuring GMPP.

Two topologies in the ANN connection are applicable: feed-forward neural networks (Figure 1.15) and recurrent neural networks (Figure 1.16). For the feed-forward ones, the input–output data are strictly feed-forward [33]. A detailed description of the data processing process can be found in Ref. [34]. However, some main differences between the two types are listed herein. The connections in feed-forward topologies between nodes in the same or previous layers are from the outputs to the inputs only. However, in recurrent topologies, it depends on dynamic characteristics. A special

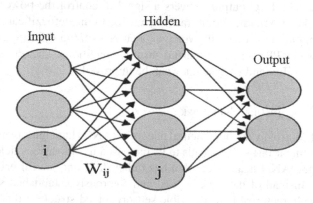

FIGURE 1.14 The three layers of an ANN structure [31].

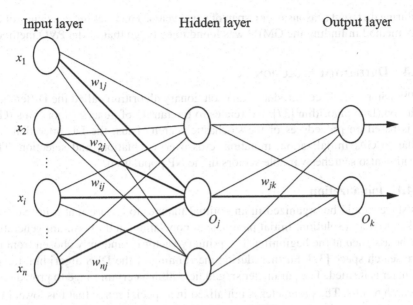

FIGURE 1.15 Multilayer feed-forward neural network [33].

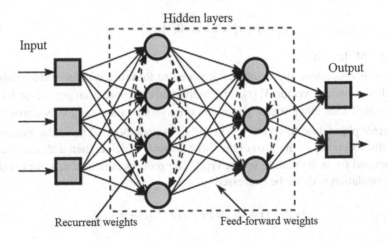

FIGURE 1.16 The recurrent neural network topology with multiple hidden layers [36].

feedback with short-term memory is present that provides a relaxation period during which the input signals don't change. This helps to consider the dynamic characteristics of ANN [34]. An ANN that combines fuzzy logic with a polar information controller has been proposed to enhance the MPPT for the PV system under partial shading conditions [35]. This work had MPPT based on a three-layer feed-forward topology ANN. The training process was fulfilled using several partially shaded PV arrays with the aim of finding the optimal PV voltage. The method developed in this work has been compared to the result obtained using the P&O, taking into consideration several shading patterns. The experimental comparison was carried out during different seasons over a span of some years. The tracking efficiency of the ANN method in finding the GMPP was found to be twice that of the P&O method.

1.2.4 DIFFERENTIAL EVOLUTION

In 1996, Storn and Rice introduced an evolutionary algorithm called the Differential Evolution (DE) algorithm [37]. DE relates to the family of genetic algorithms (GA) and is based on procedures of the stochastic search. Thus, the DE operations are similar to GA: initialization, mutation, crossover, evaluation, and selection. This algorithm also searches variable vectors in the NP population.

1.2.4.1 Initialization

This stage should be organized in an optimal manner to ensure that minimal time is taken to find a solution. Initial parameters, population, and maximum generation must be assigned at the beginning. The primary vector is randomly chosen from the entire search space [37]. Further, during the running of the DES algorithm, the jth parameter is formed. The parameter should lie within a certain range to provide better search results. The parameter is initialized in a special range that has lower $\left(x_j^L\right)$ and upper $\left(x_j^U\right)$ limits. At that time, the jth parameter in ith population is uniformly and randomly distributed between 0 and 1.

$$x_{ij}(0) = x_j^L + \mathrm{r} \text{ and } (0,1) \cdot \left(x_j^u - x_j^L\right) \tag{1.14}$$

1.2.4.2 Mutation

During this operation, all vectors are mutated to form a second variable (individuals) called a mutant vector, $\vec{V}_i(t)$. The individuals become a target vector for each generation. Three other vector parameters $\left(r_1, r_2, \text{and } r_3\right)$ are randomly chosen from the current population to create $\vec{V}_i(t)$ for the ith population. The scalar number (F) scales the difference of any two of the three vectors, which is then added to the third one obtained from the mutant vector $\vec{V}_i(t)$. The process creates a new jth parameter in the population and can be expressed as:

$$v_{ij}(t+1) = x_{r1,j}(t) + F \cdot (x_{r2,j}(t) - x_{r3,j} \dots \tag{1.15}$$

1.2.4.3 Crossover

The parent expands for the next generation during the crossover operation. The third individual produces a trial vector by combining the target vector and a mutant one. The DE algorithm consists of two different passage schemes: exponential and binomial schemes [38]. In the exponential passage, an integer n is randomly chosen among the numbers of [0, D_1] and serves as a starting point for the target vector. Another integer, L, is also chosen from the interval of [0, D_1] to represent the number of components. The trial vector can be expressed as:

$$\vec{U}_{i,j}(t) = \left[U_{i,1}(t), U_{i,2}(t), \dots, U_{i,D}(t)\right] \tag{1.16}$$

where

$$U_{i,j} = v_{i,j}(t) \text{ for } j = \langle n \rangle D, \langle n+1 \rangle D, ..., \langle n-L+1 \rangle D = x_{i,j}(t) \qquad (1.17)$$

The angular bracket $\langle \ \rangle D$ denotes a modulo function with modulus D. For any $m > 0, (L > m) = (CR)^{m-1}$. In the binomial passage, the passage is achieved on the D variables whenever a randomly picked number between 0 and 1 is within the CR value.

1.3 MPPT IMPLEMENTATION BY ELECTRONIC CIRCUITS

All algorithms for MPP search are based on DC–DC converters topology: BUCK [39–41], Boost [42–45], and BUCK-BOOST [46–49]. The SEPIC or Cuk converters are applied by considering the need to have a BUCK-BOOST converter without polarity reversal [49–51]. It is worth remembering the topology of the above-mentioned converter types.

In Table 1.2, V_O/V_S indicates the transformation coefficient as a relation of output vs. input voltages; $\Delta V_O/V_S$ is the coefficient of an output ripple. The minimal inductance (L_{min}) ensures continuous current mode (CCM) control that is much more suitable than discontinuous current mode (DCM) voltage regulation. There are three main requirements for converters:

- Fast current-voltage control;
- Stability and sustainability of the operation;
- High efficiency.

A short review on the application of different DC–DC converters can be found in Ref. [52]. The requirement to provide fast control leads to the use of the sliding mode, which is the most rapid type of control [53–56]. Sliding mode control represents a good opportunity to accelerate MPP search speed, especially in fast-changing environmental conditions. It is based on switching (or "on–off" relay control). It causes the maximally rapid change of a controlled parameter, which is observed to manage the following "on" or "off" operation. The principle of the maximization parameter alteration velocity or minimum activating time by the discrete switching of control variables was founded by a Russian mathematics group led by L.S. Pontryagin [57,58]. The specific application of the so-called "principle of maximum" for MPPT control has been studied by Azhmyakov et al. [59]. It is worth remembering that the main problem of sliding mode is the restricted control stability, which should thus be ensured by special techniques. Sustainability of control is achieved by the determination and further indication of a range. The output parameters (V, I) should remain in this range to ensure stable control [60]. Dual feedback control is another method to improve the sustainability of sliding mode control. This approach includes internal (as a rule a current of converter inductor) and external control aims to achieve MPP [53]. The scheme of this control is represented in Figure 1.17. The inner loop is responsible for the inductor current and includes the current sensor connected to the

TABLE 1.2

Main Topologies of DC–DC Converters Applicable in MPPT Design

Topology	Equivalent Circuit	Equations

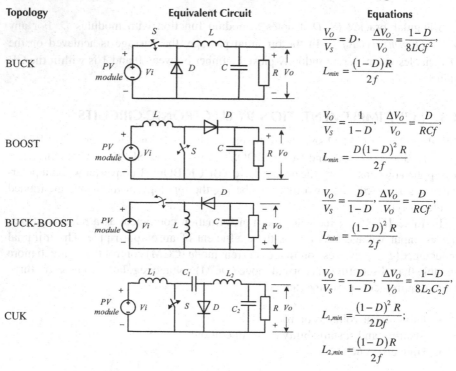

BUCK

$$\frac{V_O}{V_S} = D, \quad \frac{\Delta V_O}{V_O} = \frac{1-D}{8LCf^2},$$

$$L_{min} = \frac{(1-D)R}{2f}$$

BOOST

$$\frac{V_O}{V_S} = \frac{1}{1-D}, \quad \frac{\Delta V_O}{V_O} = \frac{D}{RCf},$$

$$L_{min} = \frac{D(1-D)^2 R}{2f}$$

BUCK-BOOST

$$\frac{V_O}{V_S} = \frac{D}{1-D}, \quad \frac{\Delta V_O}{V_O} = \frac{D}{RCf},$$

$$L_{min} = \frac{(1-D)^2 R}{2f}$$

CUK

$$\frac{V_O}{V_S} = \frac{D}{1-D}, \quad \frac{\Delta V_O}{V_O} = \frac{1-D}{8L_2 C_2 f},$$

$$L_{1,min} = \frac{(1-D)^2 R}{2Df};$$

$$L_{2,min} = \frac{(1-D)R}{2f}$$

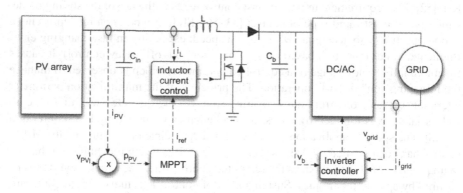

FIGURE 1.17 Block diagram of a control system [53].

comparison element. It also includes an MPPT block responsible for organizing the reference signal of a desirable current value, which ensures MPP.

A similar approach was also used previously [61–63]. Interleaved Boost DC–DC converter was used to decrease power losses caused by ripple current in the MPPT

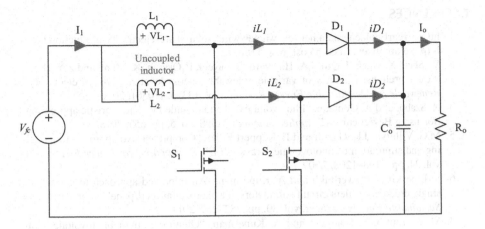

FIGURE 1.18 Interleaved Boost DC–DC converter in MPPT circuit [64].

circuit (Figure 1.18). The proposed scheme applies two parallel operating Boost converters between the PV module and the output load connected with the ripple smoothing capacitor (C). Two switching elements (S_1, S_2) of converters (as of rule transistors) are working synchronously and oppositely: If one of them is in the "on" mode, the other is in the "off" mode, and vice versa. In this way, the current ripple and power losses are diminished.

Several techniques are aimed at improving the stability of the MPPT electronic system [65–67]. The application of the Cuk DC–DC multiple switches converter allows to improve the functionality and may be efficient for stand-alone PV facilities [68]. Another type of similar SEPIC converter is also discussed for the same [69].

Different circuitry approaches are studied to decrease the MPP search time while ensuring stable control process [70–72].

1.4 CONCLUSION

This chapter provided a comprehensive overview of the MMPT algorithms and systems for PV power plants aiming to obtain maximum achievable energy. The classification of algorithms based on the analysis of their functionality, as by the deterministic or the stochastic approach, was suggested. The importance of the deterministic methods in cases of uniform solar irradiation of solar modules connected together was demonstrated. At the same time, the stochastic methods were also shown to have advantages in certain conditions such as partial shading. The proposed classification was added by the typical representatives of the mentioned methods with a detailed explanation of the functionality. The state-of-the-art MPPT algorithms and several typical topologies were presented. A detailed analysis of DC–DC converters usable in MPPT systems was also provided, emphasizing the efficiency of the different topologies.

REFERENCES

1. https://yearbook.enerdata.net/renewables/wind-solar-share-electricity-production.html
2. http://solarcellcentral.com/cost_page.html
3. W. Shen, X. Chen, J. Qiu, J.A. Hayward, S. Sayeef, P. Osman, K. Meng and Z.Y. Dong, "A comprehensive review of variable renewable energy levelized cost of electricity," *Renewable and Sustainable Energy Reviews*, vol. 133, p. 110301, 2020.
4. R. Szabo, and A. Gontean. "Photovoltaic cell and module IV characteristic approximation using Bézier curves," *Applied Sciences*, vol. 8, no. 5, pp. 655, 2018.
5. M.G. Villalva, J.R. Gazoli and E. Ruppert Filho, "Comprehensive approach to modeling and simulation of photovoltaic arrays," *IEEE Transactions on Power Electronics*, vol. 24, pp. 1198–1208, 2009.
6. S. Lineykin, M. Averbukh and A. Kuperman, "An improved approach to extract the single-diode equivalent circuit parameters of a photovoltaic cell/panel," *Renewable and Sustainable Energy Reviews*, vol. 30, pp. 282–289, 2014.
7. M. Averbukh, S. Lineykin and A. Kuperman, "Obtaining small photovoltaic array operational curves for arbitrary cell temperatures and solar irradiation densities from standard conditions data," *Progress in Photovoltaics Research and Applications*, vol. 21, no. 5, pp. 1016–1024, 2013.
8. Y. Tao, J. Bai, R.K. Pachauri and A. Sharma, "Parameter extraction of photovoltaic modules using a heuristic iterative algorithm," *Energy Conversion and Management*, vol. 224, pp. 113386, 2020.
9. S. Rajput, M. Averbukh, A. Yahalom and T. Minav, "An approval of MPPT based on PV Cell's simplified equivalent circuit during fast-shading conditions," *Electronics*, vol. 8, no. 9, pp. 1060, 2019.
10. M. Lasheen, A.K.A. Rahman, M. Abdel-Salam and S. Ookawara, "Performance enhancement of constant voltage based MPPT for photovoltaic applications using genetic algorithm," *Energy Procedia*, vol. 100, no. 100, pp. 217–222, 2016.
11. M. Lasheen, A.K.A. Rahman, M. Abdel-Salam and S. Ookawara, "Adaptive reference voltage-based MPPT technique for PV applications," *IET Renewable Power Generation*, vol. 11, no. 5, pp. 715–722, 2017.
12. A.R. Nansur, A.S.L. Hermawan and F.D. Murdianto, "Constant Voltage Control Using Fuzzy Logic Controller (FLC) to Overcome The Unstable Output Voltage of MPPT in DC Microgrid System," in *2018 International Electronics Symposium on Engineering Technology and Applications (IES-ETA)*, 2018, pp. 19–24.
13. K. Moutaki, H. Ikaouassen, A. Raddaoui and M. Rezkallah, "Lyapunov Function Based Control for Grid-Interfacing Solar Photovoltaic System with Constant Voltage MPPT Technique," in *International Conference on Advanced Intelligent Systems for Sustainable Development*, 2018, pp. 210–219.
14. N.A. Kamarzaman and C.W. Tan, "A comprehensive review of maximum power point tracking algorithms for photovoltaic systems," *Renewable and Sustainable Energy Reviews*, vol. 37, pp. 585–598, September 2014.
15. M. Kamran, M. Mudassar, M.R. Fazal, M.U. Asghar, M. Bilal and R. Asghar, "Implementation of improved perturb & observe MPPT technique with confined search space for standalone photovoltaic system," *Journal of King Saud University-Engineering Sciences*, vol. 32, no. 7, pp. 432–441, 2018.
16. V.K. Devi, K. Premkumar, A.B. Beevi and S. Ramaiyer, "A modified perturb & observe MPPT technique to tackle steady state and rapidly varying atmospheric conditions," *Solar Energy*, vol. 157, pp. 419–426, 2017.
17. M. Abdel-Salam, M. El-Mohandes and M. Goda, "An improved perturb-and-observe based MPPT method for PV systems under varying irradiation levels," *Solar Energy*, vol. 171, pp. 547–561, 2018.

18. R. Alik and A. Jusoh, "Modified perturb and observe (P&O) with checking algorithm under various solar irradiation," *Solar Energy*, vol. 148, pp. 128–139, 2017.
19. A. Ilyas, M. Ayyub, M.R. Khan, A. Jain and M.A. Husain, "Realisation of incremental conductance the MPPT algorithm for a solar photovoltaic system," *International Journal of Ambient Energy*, vol. 39, no. 8, pp. 873–884, 2018.
20. S. Necaibia, M.S. Kelaiaia, H. Labar, A. Necaibia and E.D. Castronuovo, "Enhanced auto-scaling incremental conductance MPPT method, implemented on low-cost microcontroller and SEPIC converter," *Solar Energy*, vol. 180, pp. 152–168, 2019.
21. L. Shang, H. Guo and W. Zhu, "An improved MPPT control strategy based on incremental conductance algorithm," *Protection and Control of Modern Power Systems*, vol. 5, no. 1, pp. 1–8, 2020.
22. C. Li, Y. Chen, D. Zhou, J. Liu and J. Zeng, "A high-performance adaptive incremental conductance MPPT algorithm for photovoltaic systems," *Energies*, vol. 9, no. 4, p. 288, 2016.
23. S. Rajput, M. Averbukh, A. Yahalom and T. Minav, "An approval of MPPT based on PV Cell's simplified equivalent circuit during fast-shading conditions," *Electronics*, vol. 8, no. 9, p. 1060, 2019.
24. T. Esram, J.W. Kimball, P.T. Krein, P.L. Chapman and P. Midya, "Dynamic maximum power point tracking of photovoltaic arrays using ripple correlation control," *IEEE Transactions on Power Electronics*, vol. 21, no. 5, pp. 1282–1291, 2006.
25. K.M. Tsang and W.L. Chan, "Maximum power point tracking for PV systems under partial shading conditions using current sweeping," *Energy Conversion and Management*, vol. 93, pp. 249–258, 15 March 2015.
26. F. Chekired, C. Larbes, D. Rekioua and F. Haddad, "Implementation of a MPPT fuzzy controller for photovoltaic systems on FPGA circuit," *Energy Procedia*, vol. 6, pp. 541–549, 2011.
27. A. Al Nabulsi and R. Dhaouadi, "Efficiency optimization of a DSP-based standalone PV system using fuzzy logic and dual-MPPT control," *IEEE Transactions on Industrial Informatics*, vol. 8, no. 3, pp. 573–584, 2012.
28. U. Yilmaz, A. Kircay and S. Borekci, "PV system fuzzy logic MPPT method and PI control as a charge controller," *Renewable and Sustainable Energy Reviews*, vol. 81, pp. 994–1001, 2018.
29. B. Bendib, F. Krim, H. Belmili, M. Almi and S. Boulouma, "Advanced fuzzy MPPT controller for a stand-alone PV system," *Energy Procedia*, vol. 50, no. 2014, pp. 383–392, 2014.
30. O.F. Kececioglu, "Robust control of high gain DC-DC converter using type-2 fuzzy neural network controller for MPPT," *Journal of Intelligent & Fuzzy Systems*, vol. 37, no. 1, pp. 941–951, 2019.
31. H. Boumaaraf, A. Talha, and O. Bouhali, "A three-phase NPC grid-connected inverter for photovoltaic applications using neural network MPPT," *Renewable and Sustainable Energy Reviews*, vol. 49, pp. 1171–1179, 2015.
32. S. Messalti, A. Harrag and A. Loukriz, "A new variable step size neural networks MPPT controller: Review, simulation and hardware implementation," *Renewable and Sustainable Energy Reviews*, vol. 68, pp. 221–233, 2017.
33. J. Han, M. Kamber, J. Pei, "9 - Classification: Advanced Methods," Data Mining (Third Edition), In The Morgan Kaufmann Series in Data Management Systems, Morgan Kaufmann Publishers, Massachusetts, United States, pp. 393–442, 2012.
34. A.M. Kassem, "MPPT control design and performance improvements of a PV generator powered DC motor-pump system based on artificial neural networks," *International Journal of Electrical Power & Energy Systems*, vol. 43, no. 1, pp. 90–98, 2012.
35. Syafaruddin, E. Karatepe and T. Hiyama, "Artificial neural network-polar coordinated fuzzy controller based maximum power point tracking control under partially shaded conditions," *IET Renewable Power Generation*, vol. 3, no. 2, pp. 239–253, June 2009.

36. J.A. Le, H.M. El-Askary, M. Allali and D.C. Struppa, "Application of recurrent neural networks for drought projections in California," *Atmospheric Research*, vol. 188, pp. 100–106, 2017.

37. K. Price, R.M. Storn, and J.A. Lampinen, "Differential evolution: a practical approach to global optimization," Springer Science & Business Media, Heidelberg, Germany, 2006.

38. C. Lin, A. Qing and Q. Feng, "A comparative study of crossover in differential evolution," *Journal of Heuristics*, vol. 17, pp. 675–703, 2011.

39. S.A. Kumar, M.N. Suneetha and C. Lakshminarayana, "The Simulation Study of Solar PV Coupled Synchronous Buck Converter with Lead Acid Battery Charging," in *2019 International Conference on Power Electronics Applications and Technology in Present Energy Scenario (PETPES)*, 2019, pp. 1–4.

40. T. Mrabti, M. El Ouariachi, B. Tidhaf, K. Kassmi, E. Chadli and K. Kassmi, "Regulation of Electric Power of Photovoltaic Generators with DC-DC Converter (Buck Type) and MPPT Command," in *2009 International Conference on Multimedia Computing and Systems*, 2009, pp. 322–326.

41. V. D. and V. John, "Dynamic Modeling and Analysis of Buck Converter based Solar PV Charge Controller for Improved MPPT Performance," in *2018 IEEE International Conference on Power Electronics, Drives and Energy Systems (PEDES)*, 2018, pp. 1–6.

42. S. Carreon-Bautista, A. Eladawy, A.N. Mohieldin and E. Sánchez-Sinencio, "Boost converter with dynamic input impedance matching for energy harvesting with multi-array thermoelectric generators," *IEEE Transactions on Industrial Electronics*, vol. 61, no. 10, pp. 5345–5353, 2014.

43. D. Das, S. Madichetty, B. Singh and S. Mishra, "Luenberger observer based current estimated boost converter for PV maximum power Extraction—A current sensorless approach," *IEEE Journal of Photovoltaics*, vol. 9, no. 1, pp. 278–286, 2018.

44. F. Pulvirenti, A. La Scala, D. Ragonese, K. D'Souza, G.M. Tina and S. Pennisi, "4-phase interleaved boost converter with IC controller for distributed photovoltaic systems," *IEEE Transactions on Circuits and Systems I: Regular Papers*, vol. 60, no. 11, pp. 3090–3102, 2013.

45. D. Jung, Y. Ji, S. Park, Y. Jung and C. Won, "Interleaved soft-switching boost converter for photovoltaic power-generation system," *IEEE Transactions on Power Electronics*, vol. 26, no. 4, pp. 1137–1145, 2010.

46. J. Shiau, M. Lee, Y. Wei and B. Chen, "Circuit simulation for solar power maximum power point tracking with different buck-boost converter topologies," *Energies*, vol. 7, no. 8, pp. 5027–5046, 2014.

47. T. Kok Soon, S. Mekhilef and A. Safari, "Simple and low-cost incremental conductance maximum power point tracking using buck-boost converter," *Journal of Renewable and Sustainable Energy*, vol. 5, no. 2, p. 023106, 2013.

48. K.T. Ahmed, M. Datta and N. Mohammad, "A novel two switch non-inverting buck-boost converter based maximum power point tracking system," *International Journal of Electrical and Computer Engineering*, vol. 3, no. 4, p. 467, 2013.

49. N. Kahoul and M. Mekki, "Adaptive P&O MPPT technique for photovoltaic buck-boost converter system," *International Journal of Computer Applications*, vol. 112, no. 12, pp. 23–27, 2015.

50. K. Nathan, S. Ghosh, Y. Siwakoti and T. Long, "A new DC–DC converter for photovoltaic systems: Coupled-inductors combined Cuk-SEPIC converter," *IEEE Transactions on Energy Conversion*, vol. 34, no. 1, pp. 191–201, 2018.

51. K.K. Tse, M.T. Ho, H.H. Chung and S.Y. Hui, "A novel maximum power point tracker for PV panels using switching frequency modulation," *IEEE Transactions on Power Electronics*, vol. 17, no. 6, pp. 980–989, 2002.

52. M.H. Taghvaee, M.A.M. Radzi, S.M. Moosavain, H. Hizam and M.H. Marhaban, "A current and future study on non-isolated DC–DC converters for photovoltaic applications," *Renewable and Sustainable Energy Reviews*, 17, pp. 216–227, 2013.

53. E. Bianconi, J. Calvente, R. Giral, E. Mamarelis, G. Petrone, C.A. Ramos-Paja, G. Spagnuolo and M. Vitelli, "A fast current-based MPPT technique employing sliding mode control," *IEEE Transactions on Industrial Electronics*, vol. 60, no. 3, pp. 1168–1178, 2012.

54. P.E. Kakosimos, A.G. Kladas and S.N. Manias, "Fast photovoltaic-system voltage-or current-oriented MPPT employing a predictive digital current-controlled converter," *IEEE Transactions on Industrial Electronics*, vol. 60, no. 12, pp. 5673–5685, 2012.

55. A. Kihal, F. Krim, A. Laib, B. Talbi and H. Afghoul, "An improved MPPT scheme employing adaptive integral derivative sliding mode control for photovoltaic systems under fast irradiation changes," *ISA Transactions*, vol. 87, pp. 297–306, 2019.

56. T.K. Soon and S. Mekhilef, "A fast-converging MPPT technique for photovoltaic system under fast-varying solar irradiation and load resistance," *IEEE Transactions on Industrial Informatics*, vol. 11, no. 1, pp. 176–186, 2014.

57. R.E. Kopp, "Pontryagin maximum principle," in *Mathematics in Science and Engineering*, Anonymous: Elsevier, 1962, pp. 255–279.

58. L.S. Pontryagin, Mathematical theory of optimal processes, Routledge, New York, United States, 2018.

59. V. Azhmyakov, E.I. Vemest, L.A.G. Trujillo and P.A. Valenzuela, "Optimization of affine dynamic systems evolving with state suprema: New perspectives in maximum power point tracking control," in *2017 IEEE 3rd Colombian Conference on Automatic Control (CCAC)*, 2017, pp. 1–7.

60. Y. Levron and D. Shmilovitz, "Maximum power point tracking employing sliding mode control," *IEEE Transactions on Circuits and Systems I: Regular Papers*, vol. 60, no. 3, pp. 724–732, 2013.

61. D.G. Montoya, C.A. Ramos-Paja and R. Giral, "Improved design of sliding-mode controllers based on the requirements of MPPT techniques," *IEEE Transactions on Power Electronics*, vol. 31, no. 1, pp. 235–247, 2016.

62. C. Meza, D. Biel, J. Negroni and F. Guinjoan, "Boost-buck inverter variable structure control for grid-connected photovoltaic systems with sensorless MPPT," in *Proceedings of the IEEE International Symposium on Industrial Electronics, ISIE*, 2005, pp. 657–662.

63. E. Mamarelis, G. Petrone and G. Spagnuolo, "Design of a sliding-mode-controlled SEPIC for PV MPPT applications," *IEEE Transactions on Industrial Electronics*, vol. 61, no. 7, pp. 3387–3398, 2013.

64. N. Selvaraju, P. Shanmugham and S. Somkun *"Two-Phase Interleaved Boost Converter Using Coupled Inductor for Fuel Cell Applications,"* *Energy Procedia*, vol. 138, pp. 199–204, 2017.

65. M. Sitbon, I. Aharon, M. Averbukh, D. Baimel and M. Sassonker (Elkayam), "Disturbance observer based robust voltage control of photovoltaic generator interfaced by current mode buck converter," *Energy Conversion and Management*, vol. 209, p. 112622, 2020.

66. R. Rajasekaran and P.U. Rani, "Bidirectional DC-DC converter for microgrid in energy management system," *International Journal of Electronics*, pp. 1–22, 2020.

67. L. Xian and Y. Wang, "Photovoltaic-battery hybrid power supply applied with advanced-time-sharing switching technique and discrete ripple correlation control," in *7th International Conference on Information and Automation for Sustainability*, 2014, pp. 1–6.

68. M. Dimitrijević, M.A. Stošović and V. Litovski, "An MPPT controller model for a standalone PV system," *International Journal of Electronics*, pp. 1–19, 2020.

69. R.A. Shirazi, I. Ahmad, M. Arsalan and M. Liaquat, "Integral Backstepping based MPPT controller for photo-voltaic system using SEPIC converter," in 2019 *7th International Conference on Control, Mechatronics and Automation (ICCMA)*, 2019, pp. 62–67.

70. M. Farhat, O. Barambones and L. Sbita, "A real-time implementation of novel and stable variable step size MPPT," *Energies*, vol. 13, no. 18, p. 4668, 2020.

71. R. Chinnappan, P. Logamani and R. Ramasubbu, "Fixed frequency integral sliding-mode current-controlled MPPT boost converter for two-stage PV generation system," *IET Circuits, Devices & Systems*, vol. 13, no. 6, pp. 793–805, 2019.

72. U. Yilmaz, O. Turksoy and A. Teke, "Improved MPPT method to increase accuracy and speed in photovoltaic systems under variable atmospheric conditions," *International Journal of Electrical Power & Energy Systems*, vol. 113, pp. 634–651, 2019.

2 Relay Coordination Optimization for Solar PV Integrated Grid Using the Water Cycle Optimization Algorithm

Manjaree Pandit
Madhav Institute of Technology & Science, Gwalior, India

Aayush Shrivastava and Devender Saini
University of Petroleum & Energy Studies, Dehradun, India

CONTENTS

2.1 INTRODUCTION

Directional overcurrent relays are created to safeguard distribution systems because of their excellent selectivity. Regardless of the voltage level of each power distribution system, "directional overcurrent relays" (DOCRs), due to their low cost and reliability, are used widely for short- and mid-transmission and distribution systems. DOCR is intended to identify configurations that decrease the time for operation of the relay within the safeguard zone while simultaneously providing predefined timed backup for relays in adjacent areas [1,2]. Dual (freedom) configurations are considered: "dial" also recognized as time-dial and "k", which seems to be the security factor (time overload capacity) that multiply the maximum current load for the current pick up settings, both at max current load.

Maximizing the time overload factor k within the range from about 140% to 160% is an adequate solution. Optimum synchronization of certain relays involves the selection of pickup current settings (PCS) or time dial settings (TDS) to satisfy all constraints. To meet the constraints of both relatively close-end and far-end fault currents, the relatively close- and far-end approaches are discussed in Refs. [2,3]. Overcurrent relay relays (OCRs) have been used in multi-loop networks, ring feeder networks, double-end feeder structures, or single-end parallel feeder structures [4,5]. DOCRs can also be used as primary relays or local backup relays across transmission systems for integrated sub-transmission and distribution systems [2,6]. The synchronization of the DOCR is a very important process for every design of protection-appropriate relay coordination and involves selecting the appropriate relay configuration to guarantee that the corresponding primary relays initially clear the fault throughout the protected zone and that, if they fail, the corresponding backup relays will work after a lag in the coordination period [7,6].

The relay has two settings, namely, "Plug Settings" (PS) and "Time Multiplier Settings". For each relay, the time of operation is characterized by two specific settings. But, an appropriately skilled engineer is needed to address this analytically by evaluating and determining all the bugs, system contingencies, and anomalies well in advance of operation [8,9]. Also, optimization techniques are being used to resolve this issue easily. DOCRs aim to find PS and TMS values for all the relays, thus reducing the total amount of time in operation as primary relays, and meeting all coordination constraints [9–11]. The incorporation of DGs into that same distribution network, despite the several advantages in turning the radial distribution network into such a complicated looped network, must be well managed for proper operation [12,13]. Microgrid activity has a major impact on the short-circuit currents of the network and, as a consequence, the defined protection scheme is insufficient for precise operation under these circumstances [14]. The distribution system generally appears to be a radial network, requiring only overcurrent transmission to secure the system. However, the two-way flow of fault currents through which the system must fit the Directional Overcurrent (DOCR) relay allows for the inclusion of the DGs within the whole system [2]. The inverter-based DG (electronic power control) as well as the Synchronous Generator (SG). The current is determined by the nature of the DG unit installed across the network. DGs based on inverters may not change most current fault levels, but DGs based on CSGs markedly influence the overall fault level. The effect of its DG setup was analysed as well [15–17]. From this, it could be demonstrated that CSG-based DGs have a major effect on the fault current. Fault currents are often quite different from grid-connected or shielded modes of operation of the microgrid. A recent study has been conducted on microgrids across both modes with CSG [17–20]. Work related to the design of microgrids and strategies for removing harmonics have been proposed in Refs. [17,21–23]. However, the design of a protection system that is effective, reliable, and suitable for both microgrid operating modes still is a fascinating task [17].

In recent times, techniques based on evolution and swarm intelligence are also discussed in the literature. These include differential evolution [24], "modified differential evaluation", "particle swarm optimization" [25], and "optimization learning-based" algorithms [26]. Such algorithms have achieved rapid convergence and provided an optimal global solution for relay coordination where a large number of overcurrent relays are available. Compensating device series are widely installed in electricity lines that increase the lines' power transmission capacity [26]. These devices adjust the impedance of the electrical lines and lead to a new degree of fault. Also, the interaction between new distributed power sources (DER) means a shift in grid topology and changes in power grid fault levels [26]. Current relays throughout the relay coordination problems are fixed and power failure is defined when the fault level changes in the network. An adaptive form of relay coordination is needed to coordinate the relays in the power grid whenever the system faults [26].

Also seen in the literature recently is the use of evolution-based and swarm intelligence-based techniques. These include differential evolution [24], modified differential assessment, optimization of particle swarms [25] and learning-based optimization [27]. Such algorithms have been able to achieve rapid convergence and also a globally optimal solution to the issue of relay coordination when a huge number of overcurrent relays are present. Series compensating devices are installed very widely on power lines that improve the power transmission capacity of the lines [26]. Such devices adjust the impending electrical lines and lead to a new degree of fault. The interaction of new incoming distributed energy sources (DER) also involves a shift in the topology of the grid as well as changes in the level of power grid faults [26]. Current relays are set to fixed defect levels throughout the relay coordination problems as well as power failure whenever the network fault level changes. To provide relay coordination in the power grid whenever the fault level changes in the system [26], an adaptive form of relay coordination is essential.

Distributed generation (DG) pertains to eco-friendly small to medium-sized structures located near load centres. The benefits of DGs include extending power generation, eliminating energy losses, improving power performance and reliability, and delivering the load to remote areas [28,29]. The integration of DGs into the distribution network, however, affects protective equipment function and synchronization. As a result, the safety of the distribution system has become one of the most critical problems in power systems with the implementation of DG. But, as DG penetrates the network, the current level of the fault becomes different and the structure of the radial power flow is also lost [29,30]. This leads to a loss of coordination of the relay.

The impact of DGs depends on the location of integration and the size and type of DG. In this situation, it is necessary to modify the relay setting through modifying parameters such as TMS and relay pick up value. All protecting devices deployed in the grid must be reliable and responsive, and, before installation, it is important to ensure that all devices operate properly [29–31].

Overcurrent relays are among the most frequently used protective systems since they are inexpensive but operate without the need for expensive communication equipment [32]. The adaptive protection strategy will first update data from the latest system changes, and then perform load flow or fault analysis to find the input data that the evolutionary algorithms use. The proposed concept is based on a centralized, relay-connecting computer system; this model is believed to be a trend for future growth. In this work, technical communication considerations are not examined;

however, the study and efficacy of the optimization algorithms are discussed. In the immediate future, therefore, resetting the relays could be achieved by selecting groups or sending the same values to the relays [33].

Approaches based on nature-inspired algorithms, such as grey wolf optimization (GWO), hybrid grey wolf and particle swarm optimization (GWOPSO) [34,35], genetic algorithm (GA) [25], motivated by and imitating natural biological behaviours, have quickly gained prominence in solving coordination problems [32,33,36–38]. GA was also frequently reported in literature reviews owing to its usability, reliability, and easy implementation. This algorithm for the natural selection of genes consisting of choice, reproduction, or mutation is based on evolutionary theories. In this case, current parameters for both the relay dial and the pickup were calculated.

2.2 LITERATURE REVIEW

Many algorithms, such as the GA, particle swarm optimization, GWO, water cycle algorithms (WCAs), and harmony search algorithms, [39–41] have been used in an adaptive manner. All these algorithms are inspired by natural or biological behaviours, and all algorithms are capable of successful use in the resolution of relay coordination [32,42]. GAs are the most widely used algorithms owing to their reliability, usability, and ease of implementation. This algorithm is based on the selection, crossover, and mutation theory processes.

To solve relay coordination, many algorithms are now implemented for obtaining the best relay setting. The estimation of relay settings was conducted manually before computers were involved. This estimate was completely inaccurate and time-consuming [39,43].

The test and error method was used in the 1960s to identify the best relay configuration using computers [44]. The slow convergence rate of this methodology is significantly higher as are calculations for TDS. The DOCRs coordination problem was resolved by a linear programming (LP) method at the end of the 1980s [45]. Although LP is a simple and fast technique, it requires a professional to set the minimum value and trap a local minimum value. Nonlinear (NLP) programming serves to refine all relay settings and address the problem of relay coordination issue [24]. Although NLP provides better efficiency, the initial TDS values in local minima are highly challenging and there is a high likelihood that the algorithm may get stuck [44]. Recent methods for solving the problem of nonlinear coordination have become useful for meta-heuristic optimization algorithms. Various methods for tackling the coordination problem have been developed by DOCRs, such as the Particle Swarm Optimization (PSO), which was influenced by the birds' social and cooperative exploration activities and tracking activities to meet their search requirements [39,46–48].

GA simulates Darwinian evolution principles [24,48,49], and random chromosome groups are formed by mutation and crossover of strings. The variable of the solutions is adapted to suit their best fitness and the new population is created after applying greedy selection and elitism. The entire mechanism is replicated to ensure that the best possible solution can be achieved through crossover, mutation, and breeding

[48–50]. Initialization with the random population is the primary link between GA and PSO. PSO can keep track of the situation, but, unlike GA, the location of community members can only be monitored [49]. PSO is not, in fact, the most fitting, but it mainly depends on the "construction of cooperation" between people (agents) [50]. Ant colony optimization (ACO) is another algorithm that simulates ant behaviour to find the shortest route from the colony to the food source [39,50,51].

The harmonical search (HS) algorithm was designed to find an excellent harmonic state throughout the creative processes of musical composition [46,52]. Seeker algorithms were proposed based on human memory reasoning, consideration, experience, and social learning [53]. Many optimized algorithms to address DOCR coordination exist such as firefly (FFA) algorithm (FFA) [54], black hole (BH) [55], and electromagnetic field optimization (EFO) [56]. Recently, hybrid methods have been proposed for DSPD coordination, combining features like CAS/FFA [57], GASP [58], BBOS differential evaluation (DE) [44], and BBOS [59], both of which are traditionally and nature-based. WCA is a heuristic approach influenced by natural hydrology. WCA begins with the sample of population raindrops for each test case given for the initial solution. The design variables (TDS and Ip) are used to express raindrops. For each case study, the TDS or Ip ranges are defined. The main objective criterion is used to analyse the raindrop; then, the waterways, river or sea, are identified. This method is replicated until both convergence criteria and the optimum relay configuration (TDS, Ip, CTI) have been fulfilled. The key contribution of the studies listed here could be described as the formulation of an efficient optimization process, called the modified water cycle algorithm (MWCA), to address the optimal coordinating problem of the DOCR. The MWCA optimization technique was introduced to increase the accuracy of the existing WCA [60,61].

By increasing the C value of conventionally produced WCA, which gradually impacts the balance between the exploration and exploitation stage, the technique proposed reduced the area of search to achieve the global minimum in the iteration process. The results of the proposed algorithms are evaluated based on the comparison results using the proposed algorithm to achieve significant minimization for all the primary relays subject to sequential operation during the total operational time of the relay pair. The issue of formulating the coordination of the DOCR is discussed in Section 2.3.

Section 2.1 presents the introduction where the background was described in detail. Section 2.2 provides the literature review, Section 2.3 describes the problem formulation part and modelling of optimal relay coordination, Section 2.4 presents the integration of multiple DGs on IEEE 13 node feeder, and Section 2.5 describes the traditional WCA and the proposed WCA. The simulations and the most relevant findings are described in Section 2.6. Finally, Section 2.7 properly draws the main findings in the conclusion.

This study uses the evolutionary WCA, inspired by the natural phenomena of the water cycle. The results of the proposed water cycle algorithm are validated using a benchmark. This research thus uses the water cycle evolutionary algorithm. The performance of this algorithm was assessed using the standard water cycle test systems (13 node feeding).

2.3 MODELLING OPTIMAL RELAY COORDINATION MODEL

2.3.1 OBJECTIVE FUNCTION

$$\min K = \sum_{i=1}^{n} w_i t_i \qquad (2.1)$$

The number of relays is indicated by i, and the time of relay is indicated by (t). K is the total operating time of the relays isolated from the network to the faulty area. The weight, w_i, depends and is usually set to one based on the probability that a certain failure happens in each protective area.

CTI is a time relay synchronization variable to determine the gap from the operating time intervals of the main and the backup relay. For maintaining a healthy system, this is very important. The CTI relay operated on the event of a fault is explained in Equation (2.2):

$$\Delta t = t_b - t_p \qquad (2.2)$$

$\Delta t =$ CTI (Coordination time interval).

Equation (2.3) defines the overcurrent operating time. TMS is the time multiplier settings from 0 to 2 used to set the period of operation for the relay, and b is the relay constant depending on the choice of the relay:

$$t = \text{TMS}\left(\frac{a}{M^{\alpha} - 1} + b\right) \qquad (2.3)$$

$t =$ Operating time interval of overcurrent relay
a,b, $\alpha =$ Relay constants
$M =$ Plug setting multiplier

$$M = \frac{I_{SC}}{I_{pickup}} \qquad (2.4)$$

M is the plug-setting multiplier that is very important in the relay system. The value of M ranges from 0 to 1. Equation (2.4) defines M as the short-circuit ratio and captures the relay current.

2.3.2 THE CONSTRAINT FOR THE OVERCURRENT RELAY COORDINATION MODEL

$$t_b - t_p \geq \text{CTI} \qquad (2.5)$$

where t_b and t_p are the backup relay and the main relay operating times, respectively. Equation (2.5) describes the relationship between the primary relay and the backup relay during service. In this relay model, CTI is taken as 0.2 seconds.

The operating time of the relay is the function of the current pickup and the current deviation of the relay. Based on the type of relay, the operating time is calculated by standard characteristic curves or analytical equations. The equation is defined in [62,63] as follows:

$$T_i^{Min} \leq T \leq T_i^{Max} \tag{2.6}$$

Here, T_i^{Min} and T_i^{Max} are the minimum and maximum operating times of Ith the relay, respectively:

$$TMS_i^{Min} \leq TMS_i \leq TMS_i^{Max} \tag{2.7}$$

$$Max(I_{load}^{Max}\ I_{set}^{Min}) \leq I_{set_i} \leq Min(I_{fault}^{Min}\ I_{set_i}^{Max}) \tag{2.8}$$

The least pickup current setting of the relay is the maximum I_{set}^{Min} value among the minimum available tap settings on the relay, and maximum load current $I_{set_i}^{Min}$ passes through it. Similarly, the maximum pickup current setting is chosen as less than the minimum value between the maximum available tap settings I_{fault}^{Min} on the relay and the minimum fault current I_{fault}^{Min} that passes through it. In this relay model, TMS min value is taken as 0.01 and the maximum value is taken as 2.5.

2.4 IMPLEMENTATION OF PV IN THE IEEE 13 NODE TEST FEEDER

The evaluation was carried out in six different cases including solar photovoltaics, for solar penetration inserted at bus 680, 675, and 634. The voltage was measured on each bus. For all six cases shows in Table 2.1, the solar photovoltaic penetration range was between 0 and 3.4 MW.

In Table 2.1 is provided the amount of solar PV in penetrated on-grid and the amount of active and reactive load connected along with the power grid. In this table, six cases of solar PV penetration are defined which are used to analyse the performance of optimization algorithms.

TABLE 2.1

Level of Solar PV Penetration Data

PV Penetration Level in % & Values (All Value in kW)

Total Active Power Load: 3466 kW

Total Reactive Power Load: 2102 kVAR

Cases	1	2	3	4	5	6
Penetration in %	0%	20%	40%	60%	80%	100%
Total PV Power	0	693.20	1386.4	2079.60	2772.80	3466.00
PV 1	0	173.30	346.60	519.90	693.20	866.50
PV 2	0	173.30	346.60	519.90	693.20	866.50
PV 3	0	346.60	693.20	1039.80	1386.40	1733.00

2.5   WATER CYCLE ALGORITHMS

The WCA was created and developed by observing the life cycle of water and the way
it creates rivers and sea from rainfall [25,26]. Thus, this is a creative heuristic approach
influenced by nature [25]. The optimization toolbox in MATLAB® [27] was used to
perform the WCA. Number N_{sr} is the amount of desired raindrops chosen to be sea
and rivers with the least value. The key rainfall is identified as the sea. The remaining
raindrops formed individuals of the river. Besides, some streams can flow to the sea
indirectly. In the following equation, the power in the flow and sea are calculated:

$$Ns_n = \text{round}\left(\frac{\text{Cost}_n - \text{Cost}_{N_{sr}+1}}{\sum_{n=1}^{N_{sr}} C_n} \times N_{stream}\right)$$

(2.9)

$$n = 1,2,3,\ldots n$$

In the first phase (population), N_{pop} streams are created. Then the best is picked
as the sea and N_{sr} rivers (minimum values). The minimum value of the stream is
known as the shore, among other things. But N_{sr} is the sum of the rivers and a single
sea (user-specified). The other population (N_{stream}) is referred to as streams flowing
directly into the sea or through the rivers.

Here, N_{Sn} is the number of streams that flow into the rivers and the sea. In reference [33–34], the process for exploitation and exploration is defined.

River or seawater streaming. Flowing to the sea, too. Equations (2.10) and (2.11):
New place for streams and rivers:

$$x_{stream}^{i+1} = x_{stream}^i + \text{rand} \times C\left(x_{river}^i - x_{stream}^i\right)$$

(2.10)

$$x_{river}^{i+1} = x_{river}^i + \text{rand} \times C\left(x_{sea}^i - x_{river}^i\right)$$

(2.11)

C shows that a positive constant lies between 1 and 2. Optimal conditions are only
achieved when the water flows into the sea via lakes and streams, inducing further
evaporation of the water. This condition $\left|X_{sea}^k - X_{river,i}^k\right| \le d_{max}$ is intended to ensure
that local optima cannot be fooled. The exploration intensity near the sea is given by
d_{max}. In this scenario, the best optimum solution is defined and illustrated here:

$$d_{max}^{i+1} = d_{max}^i - \frac{d_{max}^i}{K_{max}}$$

(2.12)

The rainfall method is later adapted to form streams in different areas, following the
evaporation mode as per the following equation:

The solution mainly requires that the algorithm provide the following equation:

$$X_{stream}^{new} = X^{low} + v(X_{up} - X_{low})$$

(2.13)

Equality (2.13) speeds up and/or implements the process of concentration. To this
end, Equation (2.13) is particularly used to direct streams to the sea:

$$X^{sea}_{stream} = X_{sea} + \sqrt{v} \times N\{1, N_{var}\} \qquad (2.14)$$

IIere, μ is the rate of deviation and the duration of the search zone close to the shore (suitable $\mu = 0.1$). The random number $N\{1, N_{var}\}$ is allocated.

2.6 RESULT AND DISCUSSION

The solution mainly requires that the algorithm provide a solution in a fast and finite time frame that meets the defined requirement of the model. Results for comparative analysis for the OCR coordination problem are computed using WCA, GWO, GWO-PSO, and interior point algorithm. Constrained-based optimization technique is preferred to solve this kind of problem. In this model, all the constraints are defined for the optimization of the TMS value of overcurrent relay. Relay reaction time is dependent on the TMS value.

The result obtained from the WCA is shown in Table 2.2. These are all results that are found to satisfy the constraints and can give the desired TMS value. All TMS

TABLE 2.2
TMS Value Obtained through Water Cycle Algorithms

	WCA					
PV Penetration Level	0.0%	0.20%	0.40%	0.60%	0.80%	0.100%
Case No.						
Relay No.	1	2	3	4	5	6
1	0.65	0.65	0.65	0.65	0.65	0.65
2	0.45	0.45	0.45	0.45	0.45	0.45
3	0.25	0.25	0.25	0.25	0.25	0.25
4	0.05	0.05	0.05	0.05	0.05	0.05
9	0.85	0.85	0.85	0.85001	0.85016	0.85
8	0.65	0.65	0.65	0.65	0.65001	0.65
7	0.45	0.45	0.45	0.45	0.45001	0.45
6	0.25	0.25	0.25	0.25	0.25001	0.25
5	0.05	0.05	0.05	0.05	0.05001	0.05
21	0.85	0.85	0.85015	0.85004	0.85	0.85
20	0.65	0.65	0.65	0.65	0.65	0.65
19	0.45	0.45	0.45	0.45	0.45	0.45
18	0.25	0.25	0.25	0.25	0.25	0.25
17	0.05	0.05	0.05	0.05	0.05	0.05
15	0.85	0.85362	0.8502	0.85	0.85	0.85
14	0.65	0.65335	0.65	0.65	0.65	0.65
16	0.45	0.45005	0.45	0.45	0.45	0.45
13	0.25	0.25005	0.25	0.25	0.25	0.25
12	0.05	0.05	0.05	0.05	0.05	0.05
22	2.22374	0.98166	0.96457	0.8081	0.72651	0.90539
11	2.1439	0.64376	0.57189	0.4863	0.32511	0.69351
23	1.20595	0.83692	0.31792	0.26393	0.18285	0.38633
10	0.07598	0.08195	0.07992	0.06887	0.09238	0.10778

relay values are specified for each PV penetration level and every case of PV penetration is represented in Table 2.2.

The result obtained from the grey wolf algorithm is shown in Table 2.3. These are all results found to satisfy the constraint and to be able to give the desired TMS value. All TMS relay values are specified for each PV penetration level, and c1, c2, c3, c4, c5, and c6 represent penetration cases.

The GWO is applied to the IEEE 14-bus network shown in Figure 2.1. The 13-node IEEE system consists of 23 relays. The GWO was applied to the same problem with 300 particles for 500 iterations. The GWO results are listed in Table 2.4. All the outcome of GWO is noticed to be feasible and to satisfy all the constraints.

The interior point algorithm was applied to the same problem with a stopping criteria of 500 iterations. The GWO results are listed in Table 2.5. All the outcomes of interior point algorithms were noted to be feasible and to satisfy all the constraints. This algorithm takes less iteration than the GWO and GWO-PSO algorithm.

TABLE 2.3
TMS Values Obtained through the Gray Wolf Optimization Algorithm

PV Penetration Level	GWO					
Case No.	0.0%	0.20%	0.40%	0.60%	0.80%	0.100%
Relay No.	1	2	3	4	5	6
1	0.65077	0.65033	0.65093	0.65074	0.65125	0.65056
2	0.45035	0.45026	0.45005	0.45047	0.45059	0.45046
3	0.25021	0.25009	0.25001	0.25017	0.25047	0.2502
4	0.05007	0.05003	0.05	0.05	0.05	0.05
9	0.85329	0.85247	0.85436	0.85361	0.85305	0.85465
8	0.65284	0.65207	0.65198	0.65133	0.653	0.65398
7	0.45128	0.451	0.4505	0.4506	0.45201	0.45115
6	0.25063	0.25053	0.25015	0.25026	0.25071	0.2511
5	0.05	0.05	0.05014	0.05	0.05055	0.05062
21	0.85549	0.85161	0.85338	0.8533	0.85284	0.85099
20	0.65147	0.65121	0.65204	0.65259	0.65277	0.65031
19	0.45109	0.45116	0.45139	0.45205	0.45204	0.45024
18	0.25056	0.25053	0.25023	0.25106	0.25105	0.25014
17	0.05	0.05	0.05015	0.05013	0.05023	0.05
15	0.8541	0.85305	0.85279	0.85451	0.85408	0.85586
14	0.65175	0.65282	0.6525	0.65185	0.65094	0.65043
16	0.4508	0.45124	0.45178	0.45128	0.45054	0.45036
13	0.2506	0.25061	0.25085	0.2502	0.25043	0.25027
12	0.05015	0.05026	0.05002	0.05	0.05034	0.05
R22	2.22634	0.98389	0.96754	0.80914	0.72892	0.90687
R11	2.14557	0.64477	0.57193	0.48716	0.32531	0.69468
R23	1.20735	0.83733	0.31792	0.26458	0.18303	0.38744
R10.	0.07598	0.08195	0.07985	0.0692	0.09284	0.10778

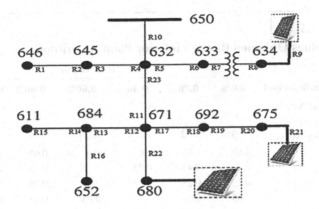

FIGURE 2.1 Solar PV implementation on IEEE 13 node feeder and representation of relay position in the grid.

TABLE 2.4
TMS Values Obtained through GWO-PSO Algorithm

	GWO-PSO					
PV Penetration Level	0.0%	0.20%	0.40%	0.60%	0.80%	0.100%
Case No.						
Relay No.	1	2	3	4	5	6
1	0.65039	0.65003	0.65504	0.65168	0.65031	0.65043
2	0.45006	0.45003	0.45245	0.45135	0.45019	0.45037
3	0.25002	0.25001	0.2521	0.25036	0.25005	0.25013
4	0.05	0.05	0.05	0.05	0.05003	0.05
9	0.85106	0.88276	0.85501	0.86965	0.85062	0.91793
8	0.65051	0.67061	0.65295	0.66031	0.65033	0.68177
7	0.45046	0.46711	0.45287	0.45418	0.45021	0.46893
6	0.25043	0.26468	0.25211	0.25205	0.25005	0.26592
5	0.05	0.05	0.05018	0.05002	0.05001	0.05
21	0.85627	0.86022	0.89558	0.88937	0.85303	0.8503
20	0.65548	0.65372	0.69155	0.67845	0.65071	0.65025
19	0.45257	0.45185	0.48666	0.46073	0.45028	0.45018
18	0.25028	0.25086	0.25436	0.25094	0.2502	0.25001
17	0.05	0.05042	0.05109	0.05001	0.05	0.05
15	0.85759	0.85149	0.90009	0.85001	0.88413	0.85152
14	0.65712	0.65064	0.68087	0.65001	0.66355	0.65085
16	0.45702	0.45004	0.46529	0.45	0.45753	0.45071
13	0.25019	0.25003	0.25944	0.25	0.25059	0.25051
12	0.05	0.05002	0.05	0.05	0.05	0.05
22	2.22421	1.00988	1.03646	0.80891	0.73016	0.9088
11	2.14452	0.66474	0.60583	0.48689	0.32664	0.69388
23	1.20639	0.84496	0.34196	0.26433	0.18305	0.38645
10	0.07598	0.08195	0.07984	0.06887	0.09238	0.10778

TABLE 2.5

TMS Values Obtained through Interior Point Algorithm

PV Penetration Level	Interior Point					
Case No.	0.0%	0.20%	0.40%	0.60%	0.80%	0.100%
Relay No.	1	2	3	4	5	6
1	0.65	0.65	0.65	0.65	0.65	0.65
2	0.45	0.45	0.45	0.45	0.45	0.45
3	0.25	0.25	0.25	0.25	0.25	0.25
4	0.05	0.05	0.05	0.05	0.05	0.05
9	0.85	0.85	0.85	0.85	0.85	0.85
8	0.65	0.65	0.65	0.65	0.65	0.65
7	0.45	0.45	0.45	0.45	0.45	0.45
6	0.25	0.25	0.25	0.25	0.25	0.25
5	0.05	0.05	0.05	0.05	0.05	0.05
21	0.85	0.85	0.85	0.85	0.85	0.85
20	0.65	0.65	0.65	0.65	0.65	0.65
19	0.45	0.45	0.45	0.45	0.45	0.45
18	0.25	0.25	0.25	0.25	0.25	0.25
17	0.05	0.05	0.05	0.05	0.05	0.05
15	0.85	0.85	0.85	0.85	0.85	0.85
14	0.65	0.65	0.65	0.65	0.65	0.65
16	0.45	0.45	0.45	0.45	0.45	0.45
13	0.25	0.25	0.25	0.25	0.25	0.25
12	0.05	0.05	0.05	0.05	0.05	0.05
22	2.22374	0.98166	0.9645	0.8081	0.72651	0.90539
11	2.1439	0.64376	0.57183	0.4863	0.32511	0.69351
23	1.20595	0.83692	0.31786	0.26393	0.18285	0.38633
10	0.07598	0.08195	0.07984	0.06887	0.09238	0.10778

By examining the results provided in Tables 2.2–2.5, we see that the modified WCA can find a better solution than the GWO, GW-PSO, and interior point algorithm with fewer iterations. The biggest drawback of using internal point techniques is that they are largely dependent on the starting point. On the other hand, heuristic techniques, using a population of random starting points solved this problem. This can be seen from the findings in Tables 2.2–2.4. Where WCA was able to find a solution that was more feasible than GWO and GWO-PSO with fewer iterations, interior point algorithm required less iteration but higher computational time than WCA. This makes the WCA suitable for online adaptive safety where the relay settings can be modified as the design of the system changes.

For each algorithm which holds at 500, maximum iteration shall be the stop criterion. Tables 2.2–2.4 show experimental results after 20 simulations, where the WCA is more effective in minimizing functions than GWO and GWOPSO. Table 2.6 provides further clarification regarding the values of the objective function or

TABLE 2.6
Relay Operation Time at Each Case and for All Different Algorithms Settings

WCA
Operation Time of Relay (seconds)

PV Penetration Level	0.0%	0.20%	0.40%	0.60%	0.80%	0.100%
Case No.						
Relay No.	1	2	3	4	5	6
1	0.65	0.65	0.65	0.65	0.65	0.65
2	0.45	0.45	0.45	0.45	0.45	0.45
3	0.25	0.25	0.25	0.25	0.25	0.25
4	0.05	0.05	0.05	0.05	0.05	0.05
9	0.85	0.85	0.85	0.85001	0.85016	0.85
8	0.65	0.65	0.65	0.65	0.65001	0.65
7	0.45	0.45	0.45	0.45	0.45001	0.45
6	0.25	0.25	0.25	0.25	0.25001	0.25
5	0.05	0.05	0.05	0.05	0.05001	0.05

GWO

Case No.						
Relay No.	1	2	3	4	5	6
1	0.65077	0.65033	0.65093	0.65074	0.65125	0.65056
2	0.45035	0.45026	0.45005	0.45047	0.45059	0.45046
3	0.25021	0.25009	0.25001	0.25017	0.25047	0.2502
4	0.05007	0.05003	0.05	0.05	0.05	0.05
9	0.85329	0.85247	0.85436	0.85361	0.85305	0.85465
8	0.65284	0.65207	0.65198	0.65133	0.653	0.65398
7	0.45128	0.451	0.4505	0.4506	0.45201	0.45115
6	0.25063	0.25053	0.25015	0.25026	0.25071	0.2511
5	0.05	0.05	0.05014	0.05	0.05055	0.05062

GWO-PSO

Case No.						
Relay No.	1	2	3	4	5	6
1	0.65039	0.65003	0.65504	0.65168	0.65031	0.65043
2	0.45006	0.45003	0.45245	0.45135	0.45019	0.45037
3	0.25002	0.25001	0.2521	0.25036	0.25005	0.25013
4	0.05	0.05	0.05	0.05	0.05003	0.05
9	0.85106	0.88276	0.85501	0.86965	0.85062	0.91793
8	0.65051	0.67061	0.65295	0.66031	0.65033	0.68177
7	0.45046	0.46711	0.45287	0.45418	0.45021	0.46893
6	0.25043	0.26468	0.25211	0.25205	0.25005	0.26592
5	0.05	0.05	0.05018	0.05002	0.05001	0.05

(Continued)

TABLE 2.6 (*Continued*)

Relay Operation Time at Each Case and for All Different Algorithms Settings

		WCA					
		Operation Time of Relay (seconds)					
PV Penetration Level		**0.0%**	**0.20%**	**0.40%**	**0.60%**	**0.80%**	**0.100%**
				Interior Point			
Case No.							
Relay No.		**1**	**2**	**3**	**4**	**5**	**6**
1		0.65	0.65	0.65	0.65	0.65	0.65
2		0.45	0.45	0.45	0.45	0.45	0.45
3		0.25	0.25	0.25	0.25	0.25	0.25
4		0.05	0.05	0.05	0.05	0.05	0.05
9		0.85	0.85	0.85	0.85	0.85	0.85
8		0.65	0.65	0.65	0.65	0.65	0.65
7		0.45	0.45	0.45	0.45	0.45	0.45
6		0.25	0.25	0.25	0.25	0.25	0.25
5		0.05	0.05	0.05	0.05	0.05	0.05

fitness function that are the overall operating time. The lowest value is selected as the best solution. The standard deviation from the mean WCA technique is less than that from the other techniques. This statistic shows that the WCA is very little different from the expected mean and its superiority is therefore also justified here (Table 2.7).

Figures 2.2 and 2.3 display the convergence graph of all algorithms and recognize that the interior point algorithm and the WCA algorithm are taken less

TABLE 2.7

Comparison of Standard Deviation Results of All Algorithms

	Mean Results of All Algorithms			
Algorithms				
Case	**WCA**	**GWO**	**GWO-PSO**	**Interior Point**
Case 1	1.401	1.422	1.411	1.401
Case 2	1.399	1.421	1.412	1.400
Case 3	1.401	1.422	1.413	1.403
Case 4	1.401	1.423	1.411	1.404
Case 5	1.402	1.426	1.413	1.405
Case 6	1.401	1.422	1.413	1.402

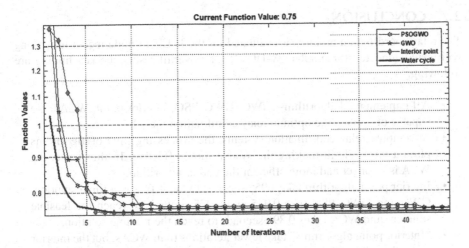

FIGURE 2.2   Convergence graph of all algorithms for PV penetration for case 4.

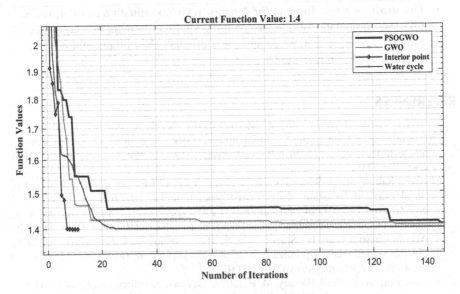

FIGURE 2.3   Convergence graph of all algorithms for PV penetration for case 6.

iteration compared to the GWO & GWO-PSO algorithm. The interior point algorithm takes 12 iterations, GWO takes 46 iterations, GWO-PSO takes 43 iterations to reach the feasible value. Figure 2.3 also shows that the interior point algorithm and the WCA algorithm require less iterations to find a feasible solution. As per the computational time taken by algorithms, the interior point algorithm takes 1.955 seconds to find the optimal solution. Similarly, GWO and GWO-PSO take 0.955 and 0.975 seconds to reach the feasible solution and WCA will take 0.896 seconds to reach this.

2.7   CONCLUSION

The relay coordination problem is evaluated for the IEEE 13 node feeder test, along with grid-connected three solar PV-DGs. In the overall result, the key finding are following:

- All implemented algorithms (GWO, GWO-PSO, interior point, WCAs) are capable of solving the optimal relay coordination model.
- In complex relay coordination systems, the processing time of algorithms plays a vital role in making the system more effective. In this sense, the WCA is stronger and more efficient than other algorithms.
- Interior point algorithm take 1.955 seconds to identify the optimal solution, GWO and GWO-PSO take 0.955 and 0.975 seconds to reach the feasible solution, and WCA takes 0.896 seconds to reach the feasible solution.
- "Interior point algorithms" take fewer iterations than WCAs, but the interior point algorithm takes high computational time to converge iterations.
- The results of the standard deviation validate that the WCA is slightly better in all applied algorithms.
- Due to all the above findings for dynamic relay coordination method, WCA take less computational time, which is the vital requirement in relay protection, thus proving that WCAs are more efficient than others and better at handling such complex models.

REFERENCES

1. Abdel-Ghany, H. A., Azmy, A. M., Elkalashy, N. I., & Rashad, E. M. (2015). Optimizing DG penetration in distribution networks concerning protection schemes and technical impact. *Electric Power Systems Research*, 128, 113–122. doi:10.1016/j.epsr.2015.07.005
2. Abdul Salam, M., & Sethulakshmi, S. (2017). Control for grid connected and intentional islanded operation of distributed generation. In *2017 Innovations in Power and Advanced Computing Technologies, i-PACT 2017* (Vol. 2017-January, pp. 1–6). Institute of Electrical and Electronics Engineers Inc. doi:10.1109/IPACT.2017.8245110
3. Abedinpourshotorban, H., Shamsuddin, S. M., Beheshti, Z., & Jawawi, N. A. (2016). Electromagnetic field optimization: A physics-inspired metaheuristic optimization algorithm. doi:10.1016/j.swevo.2015.07.002
4. Al-Roomi, A. R., & El-Hawary, M. E. (2019). Optimal coordination of double primary directional overcurrent relays using a new combinational bbo/de algorithm. *Canadian Journal of Electrical and Computer Engineering*, 42(3), 135–147. doi:10.1109/CJECE.2018.2802461
5. Alam, M. N., Das, B., & Pant, V. (2015). A comparative study of metaheuristic optimization approaches for directional overcurrent relays coordination. *Electric Power Systems Research*, 128, 39–52. doi:10.1016/j.epsr.2015.06.018
6. Albasri, F. A., Alroomi, A. R., & Talaq, J. H. (2015). Optimal coordination of directional overcurrent relays using biogeography-based optimization algorithms. *IEEE Transactions on Power Delivery*, 30(4), 1810–1820. doi:10.1109/TPWRD.2015.2406114
7. Amraee, T. (2012). Coordination of directional overcurrent relays using seeker algorithm. *IEEE Transactions on Power Delivery*, 27(3), 1415–1422. doi:10.1109/TPWRD.2012.2190107

8. Barzegari, M., Bathaee, S. M. T., & Alizadeh, M. (2010). Optimal coordination of directional overcurrent relays using harmony search algorithm. In *2010 9th Conference on Environment and Electrical Engineering, EEEIC 2010* (pp. 321–324). doi:10.1109/EEEIC.2010.5489935

9. Bedekar, Prashant P., & Bhide, S. R. (2011). Optimum coordination of overcurrent relay timing using continuous genetic algorithm. *Expert Systems with Applications, 38*(9), 11286–11292. doi:10.1016/j.eswa.2011.02.177

10. Bedekar, Prashant Prabhakar, & Bhide, S. R. (2011). Optimum coordination of directional overcurrent relays using the hybrid GA-NLP approach. *IEEE Transactions on Power Delivery, 26*(1), 109–119. doi:10.1109/TPWRD.2010.2080289

11. Benabid, R., Zellagui, M., Chaghi, A., & Boudour, M. (2014). Application of firefly algorithm for optimal directional overcurrent relays coordination in the presence of IFCL. *International Journal of Intelligent Systems and Applications, 6*(2), 44–53. doi:10.5815/ijisa.2014.02.06

12. Bhullar, S., & Ghosh, S. (2018). Optimal integration of multi distributed generation sources in radial distribution networks using a hybrid algorithm. *Energies, 11*(3). doi:10.3390/en11030628

13. Birla, D., Maheshwari, R. P., & Gupta, H. O. (2005, March 1). Time-overcurrent relay coordination: A review. *International Journal of Emerging Electric Power Systems.* Walter de Gruyter GmbH. doi:10.2202/1553-779X.1039

14. Bouchekara, H. R. E. H., Zellagui, M., & Abido, M. A. (2017). Optimal coordination of directional overcurrent relays using a modified electromagnetic field optimization algorithm. *Applied Soft Computing Journal, 54*, 267–283. doi:10.1016/j.asoc.2017.01.037

15. Chaitusaney, S., & Yokoyama, A. (2008). Prevention of reliability degradation from recloser-fuse miscoordination due to distributed generation. *IEEE Transactions on Power Delivery, 23*(4), 2545–2554. doi:10.1109/TPWRD.2007.915899

16. Chen, C. R., Lee, C. H., & Chang, C. J. (2013). Optimal overcurrent relay coordination in power distribution system using a new approach. *International Journal of Electrical Power and Energy Systems, 45*(1), 217–222. doi:10.1016/j.ijepes.2012.08.057

17. Conde, A., & Vazquez, E. (2011). Application of a proposed overcurrent relay in radial distribution networks. *Electric Power Systems Research, 81*(2), 570–579. doi:10.1016/j.epsr.2010.10.026

18. Damchi, Y., Sadeh, J., & Mashhadi, H. R. (2015). Optimal coordination of distance and directional overcurrent relays considering different network topologies. *Iranian Journal of Electrical and Electronic Engineering, 11*(3), 231–240. doi:10.22068/IJEEE.11.3.231

19. El-Khattam, W., & Sidhu, T. S. (2009). Resolving the impact of distributed renewable generation on directional overcurrent relay coordination: A case study. *IET Renewable Power Generation, 3*(4), 415–425. doi:10.1049/iet-rpg.2008.0015

20. Eskandar, H., Sadollah, A., Bahreininejad, A., & Hamdi, M. (2012). Water cycle algorithm - A novel metaheuristic optimization method for solving constrained engineering optimization problems. *Computers and Structures, 110–111*, 151–166. doi:10.1016/j.compstruc.2012.07.010

21. Gholami, A., Shekari, T., Aminifar, F., & Shahidehpour, M. (2016). Microgrid scheduling with uncertainty: The quest for resilience. *IEEE Transactions on Smart Grid, 7*(6), 2849–2858. doi:10.1109/TSG.2016.2598802

22. Guerrero, J. M., Vasquez, J. C., Matas, J., De Vicuña, L. G., & Castilla, M. (2011). Hierarchical control of droop-controlled AC and DC microgrids - A general approach toward standardization. *IEEE Transactions on Industrial Electronics, 58*(1), 158–172. doi:10.1109/TIE.2010.2066534

23. Haj-Ahmed, M. A., & Illindala, M. S. (2013). The influence of inverter-based DGs and their controllers on distribution network protection. In *Conference Record - IAS Annual Meeting (IEEE Industry Applications Society)*. doi:10.1109/IAS.2013.6682617

24. Hatamlou, A. (2013). Black hole: A new heuristic optimization approach for data clustering. *Information Sciences*, *222*, 175–184. doi:10.1016/j.ins.2012.08.023
25. Hsieh, Y. P., Chen, J. F., Liang, T. J., & Yang, L. S. (2013). Novel high step-Up DC-DC converter for distributed generation system. *IEEE Transactions on Industrial Electronics*, *60*(4), 1473–1482. doi:10.1109/TIE.2011.2107721
26. Javadian, S. A. M., Tamizkar, R., & Haghifam, M. R. (2009). A protection and reconfiguration scheme for distribution networks with DG. In *2009 IEEE Bucharest PowerTech: Innovative Ideas Toward the Electrical Grid of the Future*. doi:10.1109/PTC.2009.5282063
27. Kalage, A. A., & Ghawghawe, N. D. (2016). Optimum coordination of directional overcurrent relays using modified adaptive teaching learning based optimization algorithm. *Intelligent Industrial Systems*, *2*, 55–71. doi:10.1007/s40903-016-0038-9
28. Kamel, S., Korashy, A., Youssef, A. R., & Jurado, F. (2020). Development and application of an efficient optimizer for optimal coordination of directional overcurrent relays. *Neural Computing and Applications*, *32*(12), 8561–8583. doi:10.1007/s00521-019-04361-z
29. Khazali, A., & Kalantar, M. (2014). Optimal power flow considering fault current level constraints and fault current limiters. *International Journal of Electrical Power and Energy Systems*, *59*, 204–213. doi:10.1016/j.ijepes.2014.02.012
30. Kida, A. A., & Gallego, L. A. (2016). Optimal coordination of overcurrent relays using mixed integer linear programming. *IEEE Latin America Transactions*, *14*(3), 1289–1295. doi:10.1109/TLA.2016.7459611
31. Korashy, A., Kamel, S., Youssef, A. R., & Jurado, F. (2019). Modified water cycle algorithm for optimal direction overcurrent relays coordination. *Applied Soft Computing Journal*, *74*, 10–25. doi:10.1016/j.asoc.2018.10.020
32. Lee, H. J., Son, G., & Park, J. W. (2011). Study on wind-turbine generator system sizing considering voltage regulation and overcurrent relay coordination. *IEEE Transactions on Power Systems*, *26*(3), 1283–1293. doi:10.1109/TPWRS.2010.2091155
33. Mansour, M. M., Mekhamer, S. F., & El-Kharbawe, N. E. S. (2007). A modified particle swarm optimizer for the coordination of directional overcurrent relays. *IEEE Transactions on Power Delivery*, *22*(3), 1400–1410. doi:10.1109/TPWRD.2007.899259
34. Méndez, E., Castillo, O., Soria, J., Melin, P., & Sadollah, A. (2017). Water cycle algorithm with fuzzy logic for dynamic adaptation of parameters. In *Lecture Notes in Computer Science (including subseries Lecture Notes in Artificial Intelligence and Lecture Notes in Bioinformatics)* (Vol. 10061 LNAI, pp. 250–260). Springer Verlag. doi:10.1007/978-3-319-62434-1_21
35. Mirjalili, S., Gandomi, A. H., Mirjalili, S. Z., Saremi, S., Faris, H., & Mirjalili, S. M. (2017). Salp Swarm Algorithm: A bio-inspired optimizer for engineering design problems. *Advances in Engineering Software*, *114*, 163–191. doi:10.1016/j.advengsoft.2017.07.002
36. Mirjalili, S., Mirjalili, S. M., & Lewis, A. (2014). Grey wolf optimizer. *Advances in Engineering Software*, *69*, 46–61. doi:10.1016/j.advengsoft.2013.12.007
37. Mohammadi, R., Abyaneh, H. A., Razavi, F., Al-Dabbagh, M., & Sadeghi, S. H. H. (2010). Optimal relays coordination efficient method in interconnected power systems. *Journal of Electrical Engineering*, *61*(2), 75–83. doi:10.2478/v10187-010-0011-x
38. Norshahrani, M., Mokhlis, H., Abu Bakar, A., Jamian, J., & Sukumar, S. (2017). Progress on protection strategies to mitigate the impact of renewable distributed generation on distribution systems. *Energies*, *10*(11), 1864. doi:10.3390/en10111864
39. Ortjohann, E., Sinsukthavorn, W., Mohd, A., Lingemann, M., Hamsic, N., & Morton, D. (2009). A hierarchy control strategy of distributed generation systems. In *2009 International Conference on Clean Electrical Power, ICCEP 2009* (pp. 310–315). doi:10.1109/ICCEP.2009.5212037

40. Purwar, E., Vishwakarma, D. N., & Singh, S. P. (2016). Optimal relay coordination for grid connected variable size DG. In *2016 IEEE 6th International Conference on Power Systems, ICPS 2016*. Institute of Electrical and Electronics Engineers Inc. doi:10.1109/ICPES.2016.7584147

41. Rajput, V. N., Pandya, K. S., & Joshi, K. (2015). Optimal coordination of directional overcurrent relays using hybrid CSA-FFA method. *ECTI-CON 2015-2015 12th International Conference on Electrical Engineering/Electronics, Computer, Telecommunications and Information Technology*, 1–6. doi:10.1109/ECTICon.2015.7207044

42. Rajput, Vipul N., & Pandya, K. S. (2017a). Coordination of directional overcurrent relays in the interconnected power systems using effective tuning of harmony search algorithm. *Sustainable Computing: Informatics and Systems*, *15*, 1–15. doi:10.1016/j.suscom.2017.05.002

43. D. Solati Alkaran, M. R. Vatani, M. J. Sanjari, G. B. Gharehpetian, and M. S. Naderi, "Optimal Overcurrent Relay Coordination in Interconnected Networks by Using Fuzzy-Based GA Method," *IEEE Trans. Smart Grid*, vol. 9, no. 4, pp. 3091–3101, 2018, doi:10.1109/TSG.2016.2626393.

44. Rivas, A. E. L., & Pareja, L. A. G. (2017). Coordination of directional overcurrent relays that uses an ant colony optimization algorithm for mixed-variable optimization problems. *Conference Proceedings -2017 17th IEEE International Conference on Environment and Electrical Engineering and 2017 1st IEEE Industrial and Commercial Power Systems Europe, EEEIC / I and CPS Europe 2017*, (September). doi.org/10.1109/EEEIC.2017.7977750

45. Sadollah, A., Eskandar, H., Lee, H. M., Yoo, D. G., & Kim, J. H. (2015). Water cycle algorithm: A detailed standard code. *SoftwareX*, 5(April), 37–43. doi:10.1016/j.softx.2016.03.001

46) Sharma, A., & Panigrahi, B. K. (2016). Optimal relay coordination suitable for grid-connected and islanded operational modes of microgrid. In *12th IEEE International Conference Electronics, Energy, Environment, Communication, Computer, Control: (E3-C3), INDICON 2015* (pp. 1–6). Institute of Electrical and Electronics Engineers Inc. doi:10.1109/INDICON.2015.7443448

47. Shen, S., Lin, D., Wang, H., Hu, P., Jiang, K., Lin, D., & He, B. (2017). An adaptive protection scheme for distribution systems with DGs based on optimized thevenin equivalent parameters estimation. *IEEE Transactions on Power Delivery*, *32*(1), 411–419. doi:10.1109/TPWRD.2015.2506155

48. Shih, M. Y., Castillo Salazar, C. A., & Conde Enríquez, A. (2015a). Adaptive directional overcurrent relay coordination using ant colony optimisation. *IET Generation, Transmission and Distribution*, *9*(14), 2040–2049. doi:10.1049/iet-gtd.2015.0394

49. Shih, M. Y., Conde Enríquez, A., & Torres Treviño, L. M. (2014). On-line coordination of directional overcurrent relays: Performance evaluation among optimization algorithms. *Electric Power Systems Research*, 110, 122–132. doi:10.1016/j.epsr.2014.01.013

50. Singh, D. K., & Gupta, S. (2012a). Optimal coordination of directional overcurrent relays: A genetic algorithm approach. *2012 IEEE Students' Conference on Electrical, Electronics and Computer Science: Innovation for Humanity, SCEECS 2012*, 1–4. doi:10.1109/SCEECS.2012.6184808

51. Singh, D. K., & Gupta, S. (2012b). Use of genetic algorithms (GA) for optimal coordination of directional over current relays. In *2012 Students Conference on Engineering and Systems, SCES 2012*. doi:10.1109/SCES.2012.6199087

52. Singh, M. (2017). Protection coordination in distribution systems with and without distributed energy resources- a review. *Protection and Control of Modern Power Systems*, *2*(1), 1–17. doi:10.1186/s41601-017-0061-1

53. Singh, M., Panigrahi, B. K., & Abhyankar, A. R. (2013). Optimal coordination of directional over-current relays using Teaching Learning-Based Optimization (TLBO) algorithm. *International Journal of Electrical Power and Energy Systems*, *50*(1), 33–41. doi:10.1016/j.ijepes.2013.02.011

54. Singh, M., Panigrahi, B. K., Abhyankar, A. R., & Das, S. (2014). Optimal coordination of directional over-current relays using informative differential evolution algorithm. *Journal of Computational Science*, *5*(2), 269–276. doi:10.1016/j.jocs.2013.05.010

55. Singh, M., & Telukunta, V. (2016). Adaptive over current relay coordination algorithm for changing short circuit fault levels. *Proceedings of the 2015 IEEE Innovative Smart Grid Technologies - Asia, ISGT ASIA 2015*. doi:10.1109/ISGT-Asia.2015.7387175

56. Sortomme, E., Venkata, M., & Mitra, J. (2010). Microgrid protection using communication-assisted digital relays (pp. 1–1). Institute of Electrical and Electronics Engineers (IEEE). doi:10.1109/pes.2010.5588146

57. Srivastava, A., Tripathi, J. M., Mohanty, S. R., & Panda, B. (2016). Optimal over-current relay coordination with distributed generation using hybrid particle swarm optimization-gravitational search algorithm. *Electric Power Components and Systems*, *44*(5), 506–517. doi:10.1080/15325008.2015.1117539

58. Susilo, L., Gu, J. C., & Huang, S. K. (2013). Fault current characterization based on fuzzy algorithm for DOCR application. *Energy and Power Engineering*, *5*, 932–936. doi:10.4236/epe.2013.54B178

59. Tharakan, K. I. (2017). Optimum coordination of using Firefly and Ant Colony Optimization Algorithm, (Iccmc), *IEEE* 617–621.

60. Tjahjono, A., Anggriawan, D. ., Faizin, A. K., Priyadi, A., Pujiantara, M., Taufik, T., & Purnomo, M. H. (2017). Adaptive modified firefly algorithm for optimal coordination of overcurrent relays. *IET Generation, Transmission and Distribution*, *11*(10), 2575–2585. doi:10.1049/iet-gtd.2016.1563

61. Uthitsunthom, D., & Kulworawanichpong, T. (2010). Optimal overcurrent relay coordination using genetic algorithms. *2010 International Conference on Advances in Energy Engineering, ICAEE 2010*, 162–165. doi:10.1109/ICAEE.2010.5557589

62. Verma, R. K., Singh, B. N., & Verma, S. S. (2017). Optimal Overcurrent Relay coordination Using GA, FFA, CSA Techniques and Comparison.

63. Zamani, A., Sidhu, T., & Yazdani, A. (2010). A strategy for protection coordination in radial distribution networks with distributed generators. In *IEEE PES General Meeting, PES 2010*. doi:10.1109/PES.2010.5589655

3 Experimental Investigation of Performance of PV Array Topologies under Simulated PSCs

Karni Pratap Palawat, Vinod Kumar Yadav, and R. L. Meena
Delhi Technological University (DTU), Delhi, India

Santosh Ghosh
Kirloskar Brothers Limited, Pune, India

CONTENTS

3.1 INTRODUCTION

In order to minimize greenhouse gas emission and climate change, several measures are being implemented globally. Decarbonization of the energy sector is one of the most important ones [1]. Fossil fuel-based energy generation is not only causing global warming but polluting nature as well, which are causing irreversible and fatal impact on humans and all other forms of lives on earth. Some of those pollutants are SO_2, NO_x, PM, and Hg [2]. Literature reveals that continuous exposure to the environment filled with the above pollutants can cause severe and irreversible damage to health [3]. Second, the use of fossil fuels non-judiciously without taking into consideration its effect on the nature leads to depletion of natural resources at a very fast rate, thereby causing irreversible damage to the ecological balance as well. While bringing a technological change, a balance is required to be maintained with nature, through use of raw materials and fuel judiciously [1]. In order to formulate the guidelines for sustainable development, we may follow in the footsteps of Mahatma Gandhi, who insisted that we must act as 'trustees' of the natural resources and hence use those wisely. Mahatma also pleaded to ensure that we bequeath a healthier planet to the future generations [4].

On the other hand, electricity is one of the most effective enablers of technological and economic development of modern society, and, in the 21st century, life without electricity is unimaginable. Hence, to find a balance between the need for energy and impact on the environment, the world has resorted to a shift towards renewable energy sources with the objective of sustainable development [1,5,6]. Policymakers, researchers, and environmentalists are postulating the importance of increasing the share of generation from renewable sources across the globe. The United Nations Framework Convention on Climate Change (UNFCCC) has set a target of restricting the increase in global temperature to 1.5°C over the baseline of pre-industrial era by the end of this century. To achieve this target, the Paris agreement was accorded in 2015, in which all member nations have committed to "intended nationally determined contribution" (INDC) to reduce carbon emission. As a result of this commitment of individual nations, total installed capacity (TIC) of solar energy is rising globally at a very fast pace and has presently crossed over 650 GW; compared with all the possible renewable energy sources, this energy resource is abundantly available, and continuous innovation has made this technology most economical. In India, at present, TIC of power generation has crossed over 370 GW; 23.5% of this (87.2 GW) is non-fossil fuel-based, of which about 9.4% (34.8 GW) comes from photovoltaic technologies [7]. In the pursuit the INDC, India has taken a target of achieving 175 GW of TIC through renewable energy by 2022 [4], among which 100 GW is going to be achieved from photovoltaic energy.

Photovoltaic technology has undergone prodigious technological change over the few decades. The journey of PV technology started with a humble beginning in the middle of the 20th century as a backup power supply provider to the transponders in satellites. But with the continual innovation in technology, project financing, and project execution, currently, PV technology has emerged as the most economic energy resource and is considered as the most preferred choice for bulk power generation [8,9]. However, there are several inherent issues associated with this PV technology.

Mismatch power losses and hotspot temperature, under partial shading conditions, are some examples. Whenever PV modules are subjected to partial shading conditions, the power output of the PV array reduces significantly [10,11]. Also, *P–V* and *I–V* behaviors exhibit nonlinearity along with producing several maximum power points, called as local maximum power points (LMPP) [12]. This further aggravates various parameters such as fill factor, power mismatch losses, heating of PV array, etc. [12]. Under mismatch conditions, the PV cells get reverse-biased and are partially shaded and behave like load to the fully illuminate cells. This leads to rise in temperature and finally results in hotspot creation [13]. The elevated temperature eventually causes further reduction in power output [9] and reduction life of the PV panels [13–16]. The partial shading may also be caused by non-uniform dust accumulation on PV modules [8,17]. This undesirable environmental phenomenon and other mismatch conditions directly causes drastic reductions in efficiency and performance ratio and cause serious safety issues [18–21]; hence, they are not desirable in any project. Many researchers have presented novel methods to mitigate hotspot phenomenon and thereby reduce cell temperature. Ghosh et al. [13] have presented a hotspot mitigation circuit (HSMC), by using self-activated IGBT, which strikes a balance between the efficacy of the proposed circuit and the complexity and cost thereof. Researchers have also demonstrated that the impact of partial shading can be substantially reduced through innovative array topologies [22,23].

The present work discusses concomitant challenges associated with photovoltaic technology and experimentally investigates the comparative performance of different array topologies under partial shading condition. For experimental study, in the present work, three array topologies are considered: series-parallel (SP), bridge-link (BL), and total-cross-tied (TCT), for interconnecting 3×3 matrix of PV modules (Make: Exide India limited, Model: EIL-40). In the experimental work, different partial shading conditions that may occur on site are simulated artificially and the performance parameters are captured through real-time data acquisition, under all considered circuit configurations and partial shading conditions. The experimental work was carried out in the month of January, during which the atmosphere is found to be clean and the irradiance remained more or less constant.

Following this brief introductory section, the rest of the chapter is organized as follows. The theoretical background is presented in Section 3.2 along with the equivalent circuit model of the PV cell and the *I–V* and *P–V* characteristics under normal as well as under partial shading conditions. Section 3.3 delineates the experimental setup and the design of experimental work, followed by the result and discussion in Section 3.4. The main findings of the work are summarized in the conclusion section, *i.e.* Section 3.5.

3.2 THEORETICAL BACKGROUND

3.2.1 Equivalent Circuit Model of PV Cells

A simplified equivalent circuit model of PV cells is presented in Figure 3.1, which represents the nonlinear relationship between the electrical parameters under illuminated conditions.

As shown in Figure 3.1, the performance of an ideal PV cell, a simplified model of a PV cell, and a practical model of a PV cell vary, and the same can be represented by different segments of the equivalent circuit. The ideal model is the cornerstone of the equivalent circuit model, which basically emulates a photodiode and behaves akin to an open circuit switch under no illumination and upon bombardment of photons produces direct photocurrent between the output terminals. Hence, the equivalent circuit model as shown in Figure 2.1 is comprised of photocurrent generated by the photodiode (I_{PV}), current through the diode (I_D), a parallel resistance (R_{sh}) that represents the shunt leakage current, and series resistance (R_s) presenting the internal resistance of solar cell [13].

Voltage current equation is as follows [9,17]:

$$I_{t,cell} = I_{PV,cell} - I_D - I_{SH} \tag{3.1}$$

where the diode current is

$$I_D = I_r \left[\exp\left(\frac{V_{t,\,cell} + R_s I_{t,\,cell}}{V_{t,\,cell} \propto} \right) - 1 \right] \tag{3.2}$$

and current through the shunt resistance,

$$I_{SH} = \frac{V_{t,cell} + R_s I_{t,cell}}{R_{sh}} \quad I_{SH} = \frac{V_{t,cell} + R_s I_{t,cell}}{R_{sh}} \tag{3.3}$$

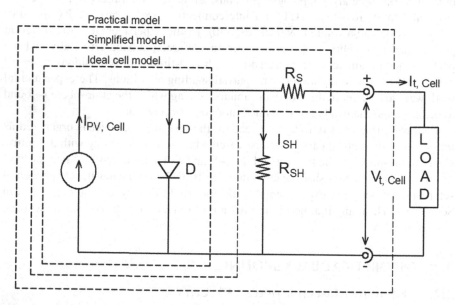

FIGURE 3.1 Equivalent circuit models of PV cell [13].

Hence, from the above three equations, by substituting Equations (3.2) and (3.3) into Equation (3.1), we get Equation (3.4) as below.

$$I_{t,cell} = I_{PV,cell} - I_r\left[\exp\left(\frac{V_{t,cell} + R_s I_{t,cell}}{V_{t,cell} \propto}\right) - 1\right] - \frac{V_{t,cell} + R_s I_{t,cell}}{R_{sh}} \quad (3.4)$$

Here, $I_{t,\ cell}$, $I_{PV,\ cell}$, and I_r are the terminal current of PV cell, current generated from the solar incident radiation, and diode reverse leakage current, respectively. $V_{t,\ cell} = k\dfrac{T}{q}k\dfrac{T}{q}$ is the terminal voltage of the PV cell, where k is Boltzmann's constant ($1.38060503 \times 10^{-33}$ J/K); T is the cell operating temperature (K); q is the charge of an electron (1.6×10^{-19} c); R_s and R_{sh} are series and shunt resistances, respectively; $\propto\propto$ is the diode emission coefficient; and the ideal value of $\propto\propto$ is 1. There are different methods to calculate the value of $\propto\propto$ to extract maximum power, as given in Refs. [26,27].

By solving the equivalent circuit model described in the paragraphs above, the nonlinear relationship between the current and power output with respect to voltage and be obtained. And from the I–V and P–V characteristics, various performance parameters as described below can be calculated.

3.3 CALCULATION OF PV PERFORMANCE PARAMETERS

3.3.1 MISMATCH POWER LOSS (ΔP_L)

$$\Delta P_L(\%) = \frac{P^{mpp} - P^{psc}}{P^{mpp}} \times 100 \quad (3.5)$$

where p^{mpp} is the maximum power output of the *PV* array, and it is calculated by multiplying v^{mpp} and I^{mpp} (refer Figure 2.2 [25]). P^{PSC} is the maximum power output produced by the PV arrays under partial shading conditions (PSC).

Efficiency (η) of any energy system is derived from the ratio of output power to input power. In the case of PV array, input is the total solar irradiance over the PV surface and the output is p^{mpp}.

Fill factor (FF) depends upon the nature of I–V characteristics and is derived as

$$FF = \frac{V^{mpp} \times I^{mpp}}{V^{oc} \times I^{sc}} \quad (3.6)$$

where v^{mpp} and I^{mpp} are voltage and current and maximum power point, and v^{oc} and I^{sc} are open circuit voltage and short circuit current, respectively (refer Figure 3.2).

Mismatch power loss is computed from the difference between global maximum power point and local maximum power, which misleads the conventional MPPT algorithms.

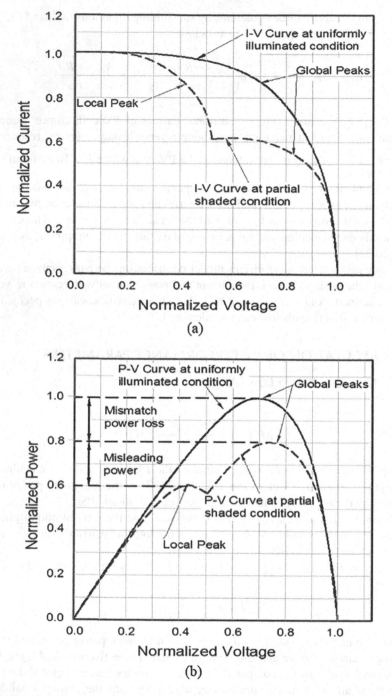

FIGURE 3.2   Characteristics of PV cells under partial shading condition: (a) *I–V* characteristics, and (b) *P–V* characteristics.

3.4 EXPERIMENTAL SETUP AND DESIGN OF EXPERIMENT

In order to experimentally investigate the relative performance of three different array topologies, a 3×3 matrix of PV modules was developed. Various partial shading patterns that may occur naturally were simulated over the array of 3×3 matrix, connected in different configurations. The scope of the experimental investigation carried out in the present work is described in Figure 3.3.

As shown in Figure 3.3, in the present work, first, three array topologies were created, which are bridge-link, series-parallel, and total cross-tied. Then five different partial shading conditions were simulated, which are (a) row shading, (b) column shading, (c) diagonal shading, (d) random shading, and (e) random shading due to tree leaves. *I–V* and *P–V* characteristics were plotted for each array configurations under all five shading patterns, and all the performance parameters, discussed in Section 3.2, were calculated. The experimental setup is illustrated in Figure 3.4. The primary results of the experiment and the derived performance parameters are presented in Section 3.4.

The instruments used for the experimental study are as follows:

a. Pyranometer: for measurement of incident solar irradiance (W/m²) on solar panels.
b. Two multi-meters: for measurement of voltage (V) and current (A).
c. Resistance/rheostat: for varying the load to trace maximum power point under PSCs and normal operation.

The technical specification of the module used for the experimental investigation is listed in Table 3.1.

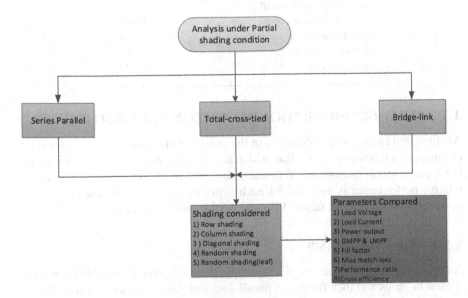

FIGURE 3.3 Scope of the experimental investigation.

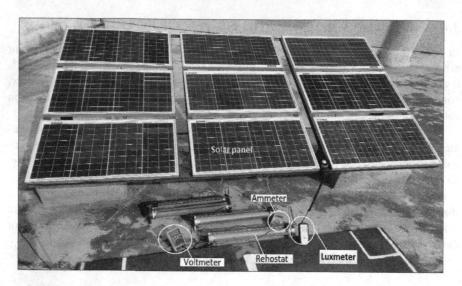

FIGURE 3.4   Experimental setup.

TABLE 3.1
Specification of the Module Used in the Study (at 1000 W/m² irradiance, AM 1.5, Temp. 25°C)

Parameter	Specified Value
Maximum power output (P_{MAX})	40 W
Voltage at maximum power (V_{MP})	18.00 V
Current at maximum power (I_{MP})	2.22 A
Short circuit current (I_{SC})	2.40 A
Open circuit voltage (V_{OC})	22.18 V

3.5   ARRAY CONFIGURATION STUDIED IN THE PRESENT WORK

As discussed in the paragraph above, in the present work, comparative performance of three array topologies are studied, which are (a) series-parallel (SP), (b) bridge-link (BL), and (c) total cross-tied (TCT) under different partial shading conditions. The relative performance is assessed for each array topology with regard to mismatch power loss, efficiency, fill factor (FF), and performance ratio (%).

3.5.1   SERIES PARALLEL (SP)

As the name itself suggests, in this topology, the 3×3 solar PV modules were connected in series and then further in parallel so that power is collected from the circuit, and schematic diagram is as shown in Figure 3.5a. This is most commonly used and the easiest and economical in application per se [22,24].

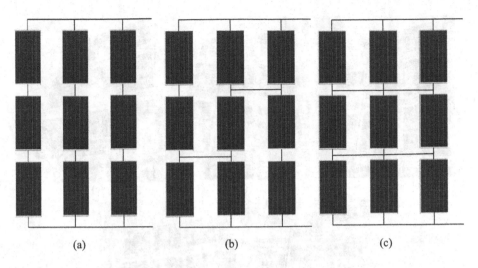

(a)   (b)   (c)

FIGURE 3.5 Array configurations: (a) series parallel, (b) bridge link, and (c) total cross tied.

3.5.2 BRIDGE-LINK (BL)

This array configuration is somewhat similar to the structure of a bridge connecting two circuits. Here, a bridge link connects two parallel paths, as shown in Figure 3.5b. The advantage is that there are more number of redundant paths than in series-parallel topology, which help in reducing mismatch power losses [25–27].

3.5.3 TOTAL CROSS-TIED (TCT)

TCT provides the maximum number of redundant paths (see Figure 3.5c), which provide alternate paths for the current to flow under partial shading conditions. This circuit configuration helps us to solve various issues associated with bridge-link (BL) and series-parallel (SP), mostly related to maximum power output under various partial shading conditions [22,26–29].

3.6 SIMULATED PARTIAL SHADING CONDITIONS

These conditions can be infinitely possible shading conditions for experimental purposes, but in this case, we have taken six partial conditions as described below and shown in Figure 3.6.

3.6.1 NO SHADING (NS)

The no shading or normal irradiance condition is considered as the first case for all the array topologies, so that the relative deterioration in performance can be quantified under different PSCs later. Under this condition, P–V plot and I–V characteristics were plotted, and all the performance parameters were calculated.

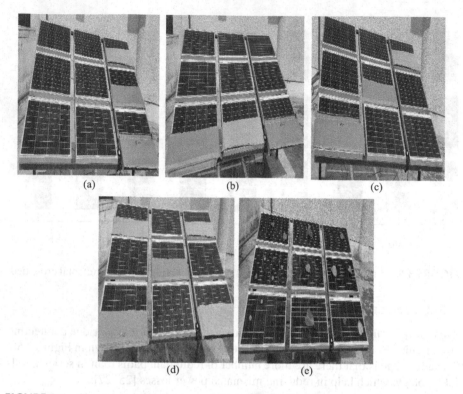

FIGURE 3.6 Simulated shading patterns : (a) column shading, (b) row shading, (c) diagonal shading, (d) random shading, and (e) random shading due to tree leaves.

3.6.2 COLUMN SHADING (CS)

As the name evinces, one of the columns of the PV array is partially covered in the experimental study to simulate column shading as shown in Figure 3.6a.

3.6.3 ROW SHADING (RS)

In this type of simulated partial shading condition, the bottom row of the PV array (see Figure 3.6b) is partially shaded. Hence, irradiance available to the PV module is relatively less as compared to the general case.

3.6.4 DIAGONAL SHADING (DS)

In this type of partial shading condition, all three modules that are diagonal are partially covered (refer Figure 3.6c) to quantify the effect on $P–V$ and $I–V$ characteristics.

3.6.5 RANDOM SHADING CONDITION

In some cases, the partial shading phenomenon occurs in a random pattern due to non-uniform dust deposition or falling leaves. In the present experimental work, two such patterns were simulated, which are random shading condition (RSC) and random shading condition due to leaves (RSL), which are illustrated in Figure 3.6d and e, respectively.

3.7   RESULT AND DISCUSSION

As discussed in Section 3.3, three array topologies were simulated in the present experimental work. All of these three topologies were tested under no shading (*i.e.* uniform irradiance) as well as under different simulated partial shading conditions. *I–V* and *P–V* characteristics were plotted for all the array topologies and under all the shading patterns. The primary results of the experimental work are the *I–V* and *P–V* curves.

Figures 3.7 and 3.8 present the *I–V* and *P–V* curves, respectively, for series-parallel array under all the partial shading conditions. It may be observed from Figures 3.7 and 3.8 that multiple peaks were introduced in both characteristics under all the simulated partial shading conditions except for row shading. The measured characteristics for no shading (uniform irradiance) are also juxtaposed for visual appreciation of the reduction of power output under partial shading conditions. The *I–V* and *P–V* curves for bridge-link array are presented in Figures 3.9 and 3.10, respectively. It will be interesting to note, from these figures, that in the case of bridge-link array, no local peaks were observed under both the random shading patterns as, on addition of the row shading pattern, which is a divergence from the observations made with regard to SP array topology. In case of total-cross-tied array (refer Figures 3.11 and 3.12), no local peaks were observed only under two shading patterns, which are random

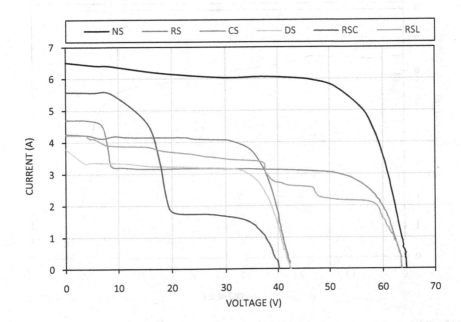

FIGURE 3.7   *I–V* characteristics of SP array under all simulated PSCs.

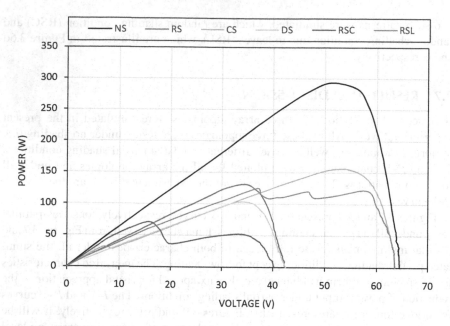

FIGURE 3.8 *P–V* characteristics of SP array under all simulated PSCs.

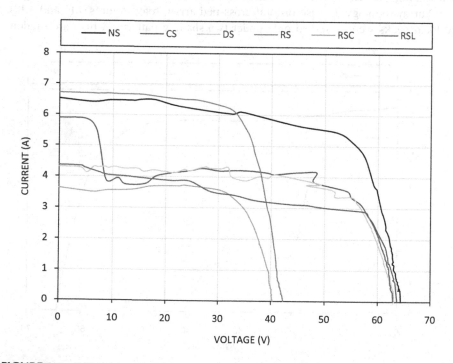

FIGURE 3.9 *I–V* characteristics of BL array under all simulated PSCs.

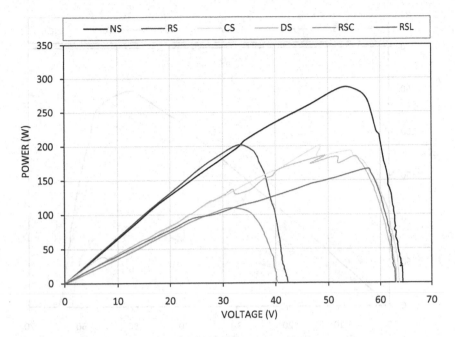

FIGURE 3.10 *P–V* characteristics of BL array under all simulated PSCs.

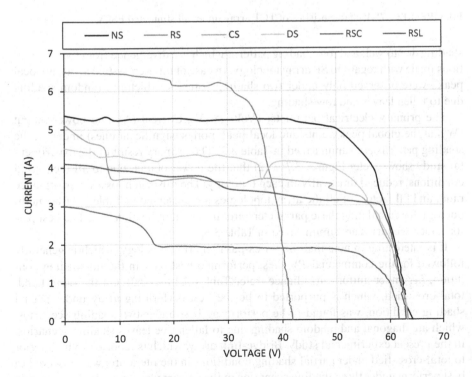

FIGURE 3.11 *I–V* characteristics of TCT array under all simulated PSCs.

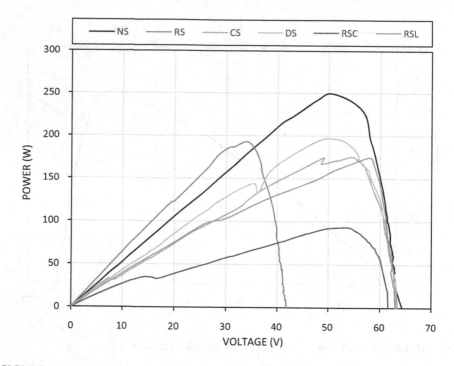

FIGURE 3.12 *P–V* characteristics of TCT array under all simulated PSCs.

shading due to well and row shading pattern, which is a divergence from the observations made with regard to SP array topology. In case of total-cross-tied array, no local peaks were observed only under two shading patterns, which are random shading due to fallen leaves and row shading.

The primary electrical parameter (Voltage (V), Current (A), and power output (W)) at the global peak points and local peak points (significant ones) under all the shading patterns are summarized in Table 3.2. The primary result of the experimental study shows (refer Figures 3.7–3.12) that the power output, due to all the partial conditions, reduced drastically in each topology. The mismatch losses, performance ratio, and fill factors of all the array topologies are presented in Table 3.2 (the methodology for calculating these parameters are discussed in Section 3.2). The best performance scenarios are summarized in Table 3.3.

It is interesting to note that the series-parallel array topology, which is generally followed for the commercial *PV* sites, performed best only in the no-shading condition, *i.e.* under uniform irradiance (refer Tables 3.2 and 3.3). On the other hand, total-cross-tied, which is purported to be the best performing array under partial shading condition, was found to be performing best under two shading scenarios, which are diagonal and random shading due to fallen tree leaves shading scenarios, in the present experimental study. Bridge-link array, which is reported to be inferior to total-cross-tied under partial shading conditions in the literature, was observed to best perform under three shading scenarios in the present work: row shading, column shading, and random shading scenarios (refer Table 3.4).

TABLE 3.2
Global and Local Maxima Power Point

Topology	PSC	GMPP			LMPP 1			LMPP 2			LMPP 3		
		V(V)	I(A)	P(W)	V(V)	I(A)	P(W)	V(V)	I(A)	P(W)	V(V)	I(A)	P(W)
SP	NS	51.2	5.73	293.38	-	-	-	-	-	-	-	-	-
	RS	34.6	3.79	131.13	-	-	-	-	-	-	-	-	-
	CS	53.9	2.9	156.31	-	-	-	-	-	-	-	-	-
	DS	33	3.12	102.96	-	-	-	-	-	-	-	-	-
	RSC	15.41	4.6	70.866	32.44	1.61	52.23	-	-	-	-	-	-
	RSL	37.04	3.37	124.82	46.7	2.57	120.02	58	2.1	121.8	-	-	-
BL	NS	52.4	5.45	285.03	-	-	-	-	-	-	-	-	-
	RS	32.9	6.13	201.68	-	-	-	-	-	-	-	-	-
	CS	48.5	4.15	201.27	54.2	3.56	192.95	-	-	-	-	-	-
	DS	55.1	3.36	185.13	32.3	4.03	130.17	55.1	3.36	185.1	51.9	3.54	183.7
	RSC	31.5	3.49	109.93	-	-	-	-	-	-	-	-	-
	RSL	57.7	2.89	166.75	-	-	-	-	-	-	-	-	-
TCT	NS	50.3	4.97	249.99	-	-	-	-	-	-	-	-	-
	RS	33.9	5.69	192.89	-	-	-	-	-	-	-	-	-
	CS	54.3	3.23	175.39	-	-	-	-	-	-	-	-	-
	DS	50.6	3.89	196.83	35.4	4.07	144.08	-	-	-	-	-	-
	RSC	53.1	1.76	93.45	14.5	2.42	35.09	-	-	-	-	-	-
	RSL	58.0	3.01	174.58	-	-	-	-	-	-	-	-	-

TABLE 3.3
Parameter Analysis of All Three Topologies

Topology	PSC	V_{MPP}	I_{MPP}	P_{MPP}	V_{OC}	I_{SC}	FF	% PR	Missmatch Power Loss (%)
SP	NS	51.2	5.73	293.38	64.5	6.51	0.69	100	0
	RS	34.6	3.79	131.13	42.3	4.26	0.73	44.69	55.30
	CS	53.9	2.9	156.31	63.5	4.7	0.52	53.27	46.72
	DS	33	3.12	102.96	42.5	3.8	0.64	35.09	64.90
	RSC	32.44	1.61	52.23	40.2	5.57	0.23	17.80	82.19
	RSL	37.04	3.37	124.82	63.6	4.23	0.46	42.54	57.45
BL	NS	52.3	5.45	285.03	64.4	6.52	0.68	100	0
	RS	32.9	6.13	201.68	42.3	6.71	0.71	70.75	29.24
	CS	48.5	4.15	201.27	63.7	5.88	0.54	70.61	29.38
	DS	55.1	3.36	185.13	63.1	4.3	0.68	64.95	35.04
	RSC	31.5	3.49	109.93	40.3	3.63	0.75	38.56	61.43
	RSL	57.7	2.89	166.75	62.9	4.36	0.61	58.50	41.49
TCT	NS	50.3	4.97	249.99	64.3	5.36	0.72	100	0
	RS	33.9	5.69	192.89	41.8	6.5	0.71	77.15	22.84
	CS	54.3	3.23	175.39	63.4	5.12	0.54	70.15	29.84
	DS	50.6	3.89	196.83	63.1	4.5	0.69	78.73	21.26
	RSC	53.1	1.76	93.45	61.6	3	0.50	37.38	62.61
	RSL	58.0	3.01	174.58	63.0	4.1	0.67	69.83	30.16

TABLE 3.4

Best Performing Array Topology under Different PSCs Studied

Partial Shading Condition	Best Topology	Maximum Power Point (GMPP) (W)
Non-shaded (NS)	Series parallel	293.376
Row shading (RS)	Bridge link	201.677
Column shading (CS)	Bridge link	201.275
Diagonal shading (DS)	Total cross tied	196.834
Random shading cardboard (RSC)	Bridge link	109.935
Random shading leaves (RSL)	Total cross tied	174.58

3.8   CONCLUSION

In the present work, the relative performances of different array topologies were quantified under different partial shading conditions, through experimental investigation. Partial shading conditions affect the *PV* performance and reliability severely due to mismatch losses and hotspot generation, respectively. This condition may be caused by non-uniform dust deposition, fallen tree leaves, bird droppings, etc., and the shading patterns in nature are too complex to accurately predict the site. Hence, this phenomenon has been studied extensively in the present work, at the laboratory scale. The deep insights provided by this experimental study are presented in this chapter.

The result of the experimental study shows that, under uniform shading scenario (*i.e.* no shading) series-parallel topology performed best, *i.e.* produced a maximum power of 293.376 W, followed by bridge link (285.035 W), while the total cross tied topology performed the worst (249.991 W). This shows that as the number of interconnecting cables and joints increases in the array, the power output will reduce due to additional losses in the cables and the joints. However, these additional cables and joints provided the redundant path for the photocurrent to flow under partial shading conditions, without reverse biasing the shaded modules and without causing a reverse breakdown of those. Hence, under partial shading conditions, the array topology having a more number of redundant paths performed better.

However, as evident from this study, the relation between the reductions in mismatch loss was not linear with the number of redundant paths and depended upon partial shading patterns. Hence, site-specific studies need to be carried out for choosing the optimal array topology in order to maximize the performance ratio of the projects, based on the detailed study of the expected shading patterns over a long period of time.

REFERENCES

1. S. Ghosh, V.K. Yadav, G. Mehta, V. Mukherjee, R. Birajdar, Status check: Journey of India's energy sustainability through renewable sources, *IFAC-PapersOnLine* 48 30 (2015) 456–461.
2. CEEW Report, India's Energy Transition: Subsidies for Fossil Fuels and Renewable Energy 2018 Update, 2018, https://www.ceew.in/publications/indias-energy-transition.

3. T. Boningari, P.G. Smirniotis, Impact of nitrogen oxides on the environment and human health : Mn-based materials for the NOx abatement, *Curr. Opin. Chem. Eng.* 13 (2016) 133–141.

4. Union Environment Ministry, India's Intended Nationally Determined Contribution, Unfccc/Indc. (2015) 1–38.

5. S. Ghosh, V. K. Yadav, G. Mehta, R. Birajdar, Evaluation of Indian power sector reform strategies and improvement direction though DEA, *IEEE Power & Energy Society General Meeting*, Chicago, IL, 2017, pp. 1–5,

6. M. K. Verma, V. Mukherjee, V. K. Yadav, S. Ghosh, Indian power distribution sector reforms: A critical review, *Energy Policy* 144 111672 Sept. 2020.

7. Central Electricity Authority (CEA), Monthly Report (June 2020)-Installed capacity. Central Electricity Authority, Ministry of Power, India, 2020. http://www.cea.nic.in/reports/ monthly/installedcapacity/2020/installed_capacity-06.pdf (accessed 02.08.2020)

8. S. Ghosh, V.K. Yadav, V. Mukherjee, Impact of environmental factors on photovoltaic performance and their mitigation strategies–A holistic review, *Renew. Energy Focus.* 28 (2019) 153–172.

9. S. Ghosh, V.K. Yadav, V. Mukherjee, Improvement of partial shading resilience of PV array though modified bypass arrangement, *Renew. Energy.* 143 (2019) 1079–1093.

10. A. Singla, K. Singh, V.K. Yadav, Environmental effects on performance of solar photovoltaic module, *2016-Biennial International Conference Power Energy System towards Sustainable Energy, PESTSE*, 2016.

11. S. Ghosh, V.K. Yadav, V. Mukherjee, P. Yadav, Evaluation of relative impact of aerosols on photovoltaic cells through combined Shannon's entropy and Data Envelopment Analysis (DEA), *Renew. Energy.* 105 (2017) 344–353.

12. Mishra, A.S. Yadav, R. Pachauri, Y.K. Chauhan, V.K. Yadav, Performance enhancement of PV system using proposed array topologies under various shadow patterns, *Sol. Energy.* 157 (2017) 641–656.

13. S. Ghosh, V.K. Yadav, V. Mukherjee, A novel hot spot mitigation circuit for improved reliability of PV module, *IEEE Trans. Device Mater. Reliab.* 20 (2020) 191–198. doi:10.1109/TDMR.2020.2970163.

14. M. Balato, L. Costanzo, M. Vitelli, Series-Parallel PV array re-configuration: Maximization of the extraction of energy and much more, *Appl. Energy.* 159 (2015) 145–160.

15. O. Kunz, R.J. Evans, M.K. Juhl, T. Trupke, Understanding partial shading effects in shingled PV modules, *Sol. Energy.* 202 (2020) 420–428.

16. Y. Hu, S. Member, J. Zhang, P. Li, D. Yu, L. Jiang, Non - uniform aged modules reconfiguration for large scale PV array, 4388 (2017) 1–9. doi:10.1109/TDMR.2017.2731850.

17. S. Mekhilef, R. Saidur, M. Kamalisarvestani, Effect of dust, humidity and air velocity on efficiency of photovoltaic cells, *Renew. Sustain. Energy Rev.* 16 (2012) 2920–2925.

18. S. Oprea, A. Bâra, D. Preoţescu, L. Elefterescu, Photovoltaic power plants (PV-PP) reliability indicators for improving operation and maintenance activities. A case study of PV-PP Agigea located in Romania, *IEEE Access* 7 (2019) 39142–39157.

19. S. De León-aldaco, H. Calleja, J. Aguayo, Reliability and mission profiles of photovoltaic systems : A FIDES Approach, *IEEE Trans. Power Electron.* 30 (2014).

20. M. Theristis, S. Member, I.A. Papazoglou, Markovian reliability analysis of standalone photovoltaic systems incorporating repairs, *IEEE J. Photovoltaics.* 4 (2014) 414–422.

21. J. Oh, B. Rammohan, A. Pavgi, S. Tatapudi, G. Tamizhmani, G. Kelly, M. Bolen, Reduction of PV module temperature using thermally conductive backsheets, *IEEE J. Photovoltaics.* 8 (2018) 1–8.

22. S. Ghosh, V. K. Yadav, V. Mukherjee, Evaluation of cumulative impact of partial shading and aerosols on different PV array topologies through combined Shannon's entropy and DEA, *Energy* 144 (2018) 765–775.

23. A.S. Yadav, V.K. Yadav, V. Mukherjee, S. Ghosh (2021) Performance investigation of different bypass diode topology based SDK-PV arrays under partial shading conditions. In: Favorskaya M., Mekhilef S., Pandey R., Singh N. (eds) *Innovations in Electrical and Electronic Engineering*. Lecture Notes in Electrical Engineering, vol 661. Springer, Singapore.

24. A. Mäki, S. Valkealahti, J. Leppäaho, Operation of series -connected silicon-based photovoltaic modules under partial shading conditions, *Prog. Photovoltaics* 20 (2012) 298–309.

25. G. Walker, Evaluating MPPT converter topologies using a matlab PV model, *J. Electr. Electron. Eng. Aust.* 21 (2001) 49–55.

26. L.F.L. Villa, D. Picault, B. Raison, S. Bacha, A. Labonne, Maximizing the power output of partially shaded photovoltaic plants through optimization of the interconnections among its modules, *IEEE J. Photovoltaics.* 2 (2012) 154–163.

27. N.K. Gautam, N.D. Kaushika, Reliability evaluation of solar photovoltaic arrays, *Sol. Energy.* 72 (2002) 129–141.

28. S. Mohammadnejad, A. Khalafi, S.M. Ahmadi, Mathematical analysis of total-cross-tied photovoltaic array under partial shading condition and its comparison with other configurations, *Sol. Energy.* 133 (2016) 501–511.

29. M.M.A. Salama, Optimal photovoltaic array reconfiguration to reduce partial shading losses, 4 (2013) 145–153.

4 Artificial Intelligence in PV System

Peeyush Kala
Women's Institute of Technology, Dehradun, India

CONTENTS

4.1   INTRODUCTION

In recent years, the penetration of renewable energy (RE) resources in power systems has increased at a rapid pace due to various factors including rise in global energy demand, fast depletion of conventional resources, and environmental concerns. Among these RE resources, solar energy holds the major share in global energy owing to its widespread availability throughout the year. Recent advancements in semiconductor technologies and power electronics have paved the way for installation of PV systems for power generation in various standalone/grid-connected residential and industrial applications. Some of the key advantages of PV systems are that they are a clean form of energy, available all around the globe, with a low maintenance and operational cost, need no rotational parts except a sun tracker system, and are easy to install. Despite various advantages of PV systems, there are various issues or challenges faced in PV systems [1]. These issues are as follows:

1. Forecasting of solar irradiance and output power
2. Improvement in efficiency and performance
3. Evaluation and assessment of power and performance
4. Fault detection

5. Prediction of optimum characteristics
6. Optimum deployment of PV arrays in power system
7. Optimal design, sizing, and control of PV systems
8. Demand side management

In literature, many researchers have proposed the use of artificial intelligence (AI) in solving the above-mentioned problems of PV systems. AI is described as an effort of humans to train and design a machine that may possess the abilities of human beings, such as thinking, learning, memorizing, pattern recognizing, taking decisions, and understanding [2]. Some of the key features of AI are as follows:

1. Pattern recognition
2. Prediction of values for the missing/noisy data
3. Ability to utilize the experience of an expert in solving problems
4. Faster and better predictions as compared to conventional methods
5. Ability to solve difficult and ill-defined problems
6. Expertise of humans can be transformed into software
7. High-speed generalization, flexible approach, and excellent prediction
8. Cognitive computing
9. Ease in deep analysis of data through simulation

The applications of AI are found in many electrical and electronics engineering problems, especially in RE system where resources exhibit intermittent behaviour. In PV systems, AI techniques have proven their decisive role in solving many reported problems. Some of these problems will be discussed in this chapter. AI techniques can accurately forecast solar irradiance. It is known that the output power of a PV system depends upon atmospheric conditions, i.e. ambient temperature, solar irradiance, wind velocity, and shading. In terms of time interval, forecasting can be classified into four types:

1. Ultra-short term
2. Short term
3. Medium term
4. Long term

Long-term forecasting is required for the installation of a PV power plant in an area, whereas for operation of PV systems, the other three modes of forecasting are required. AI can predict and provide corrective measures to improve the behaviour of PV systems based on atmospheric conditions and variations in load demand. Detection of faults and its location in PV systems is another challenge. Faults can be line-line (L-L) faults, open circuit (OC) faults, partial shading, and arc faults. L-L faults may occur when two points having different potential in a PV array get short-circuited. It is very difficult to detect faults with low mismatch when PV systems are subjected to low irradiances. This is due to the low magnitude of electrical quantities [3]. Further, use of maximum power point tracking (MPPT) also creates difficulty in the detection of LL fault. In OC faults, disconnection occurs within a string or between two adjacent strings of a PV array due to broken connecting cable or loose connection of cables. Another challenge in PV systems is to detect partial shading. It is a situation in which some of the panels of an array are covered by dust, shadows of trees or buildings, bird droppings, etc. It can be static or dynamic. Partial shading causes variation in output power and leads to many local maximum power points (MPP). Hence, in order to obtain global MPP, reconfiguration of the array is required. Apart from these faults, PV systems are also subject to arc faults when there is a failure of insulation in cables. These issues are also being tackled by AI [4].

Apart from these major problems of PV systems, AI is finding applications in optimal sizing, placing, design, and control of PV systems. To bring down the installation, operational, and maintenance cost of PV systems while achieving improvement in efficiency and performance simultaneously has become the motive of AI techniques. In this chapter, contributions of various researchers in applications of AI in PV systems is being presented and discussed in a detailed manner.

4.2   AI TECHNIQUES USED IN PV SYSTEMS

4.2.1   ENSEMBLE LEARNING

It is a type of machine learning methodology designed to improve the ability of generalization in learning models using several learners. This learning scheme is suitable for small data models. It comprises steps like boosting, bagging, and stacking. In boosting and bagging, homogeneous learners are involved, whereas heterogeneous learners are involved in stacking stage. In Ref. [5], the various ensemble-learning approaches are applied to calculate the efficiency of power conversion of organic dye-sensitized solar cells. This methodology exhibited its capability in the exploration of complex quantifiable structure activity connection for the case when features are distant from targets. Ensemble methodology outperformed methods such as support vector machine (SVM) and single base learner method and was able to achieve excellent generalization and higher accuracy. In Ref. [6], data-based ensemble methods have been used in predicting the generation of solar energy. This methodology has shown ability to handle the intermittent behaviour of generation due to PV systems. A one-day-ahead forecast of solar power generation was improved using optimized an Artificial Neural Network (ANN)-based ensemble method. Bagging and trial–error process was used for the optimization of hidden neurons in ANN model. Further, the bootstrap method was embedded into the ensemble for the estimation of the uncertainty in sources that may affect the predictions of models. A real case study was carried out on a grid-tied PV system (231 kW) for forecasting and showed that the ensemble approach-based model outclasses other benchmark models such as persistence model and optimum ANN model. The benefits of the proposed scheme were accurate short-term prediction of power generation, efficient scheduling, operational contribution of power generation in energy mix, trustworthy operation, and economic benefits. In Ref. [7], an improvised form of ensemble learning aimed at improving the forecast of power generation from the PV system is presented. The proposed ensemble model employed a combination of an adaptive residual compensation scheme and an evolutionary algorithm.

The experimental results showed that the proposed technique performed better than other conventional techniques such as least-square boosting ensemble and weighted average ensemble techniques. In PV systems, accurate forecasting of solar power is essential. There are various data-based methods like ANN, SVM, learning machines, boosting regression trees, etc. proposed for the accurate prediction of solar power generation. However, each of these methods predicts with different accuracies. Therefore, in order to find the accurate prediction, ensemble method can combine the predictions of different models and can find the mean while enhancing the accuracy of predictions.

4.2.2 DEEP LEARNING

It is termed as one of the best technique among machine learning techniques and finds use in the various following applications:

- Processing of audio or speech
- Machine translation
- Computer vision
- Filtering contents of social network
- Big data applications
- Solution of AI-based problems

In renewable energy systems (RES), it is desired to develop an accurate prediction model. AI techniques have proven their role as a tool for optimizing and predicting the power generation in RES. The challenges found in RES are analysis of large data and processing of larger number of variables. In order to tackle these problems, use of deep learning (DL) algorithms is recommended by researchers. Some of the proposed DL techniques in recent years are, namely, convolution neural network, Boltzmann machine, and auto encoder.

4.2.3 CONVOLUTIONAL NEURAL NETWORK (CNN)

This methodology consists of various layers of neurons that are trained to achieve a high performance. It comprises training in two stages, namely feed-forward and backpropagation. In the beginning, convolution operations were performed on input targets and the parameters of each neuron. The output of network is then compared with loss function, which yields error. This computed error acts as an input for next stage, i.e. backpropagation. At this stage, the parameter's gradient is computed and each parameter is altered and fed-forward. After some iterations, the training of the network ends.

4.2.4 BOLTZMANN MACHINE (BM)

These methods are based on stochastic ANN used to train probability distribution. Restricted BM and deep BM are popular among BM techniques. In restricted BM, a limitation is introduced by constructing a bipartite plot using hidden and visible units. Both of these units are independent, with some constraint. This feature offers the evolution of optimized algorithms for training. In deep BMs, there are several hidden neuron layers with individual layers of even number. During the training of layers in an unsupervised model, self-managed learning (SML) algorithm is employed to optimize the probability distribution. In literature, survey was done on deep BM technique for predicting the output of a solar farm in Germany. Data was collected for 1000 days (approx.) with a resolution of 3 hours per day.

4.2.5 Auto Encoder

Auto-encoder DL techniques incorporate the learning of encoders, which can reconstruct the input, and the corresponding output vectors also have dimensions of the input vector [8]. Optimization of encoders is done to keep the reconstruction error at a minimum. From the previous encoders, a code is learned and sent to the next auto-encoders, which are trained using backpropagation. This method is quite efficient among DL techniques.

In Ref. [9], various DL techniques implemented in PV systems are reviewed. They also proposed a new taxonomy for evaluation of the performance of PV systems. The main objectives of the review were as follows:

   a. Investigation of energy policy used with AI methodologies in PV systems
   b. Analysis of hybrid and single DL
   c. Comparison between single DL/hybrid DL with other computational intelligence methods

On the basis of this review, the optimization techniques-based DL methods have been recommended for the prediction of parameters in PV systems.

4.2.6 Machine Learning

Learning search algorithm (LSA) is inspired by the learning behaviour of disciples. The algorithm begins with a group of random learners. At this stage, students are classified among groups, namely best learner, worst learners, and average learners, on the basis of their knowledge. In the next stage of positive learning, the learner groups increase their knowledge depth. In the final stage, learners are subjected to a negative learning pattern where the best attributes of the worst learners are learned by the other learners. In this manner, the knowledge level of entire group can be enhanced. Support vector machine (SVM) is also a promising machine learning technique aimed at minimization of reconstruction error and maximization of separate margin for various classes. It is widely employed in supervised learning as a regression tool.

In Ref. [10], an ML technique was adopted for the evaluation of output electrical energy of PV inverters using five-year data under both partial shading and unshaded conditions. Prediction of the generated energy of PV systems while using long-term data analysis through the ML technique was done in this work. It was concluded that the average degradation of microinverters was at 3% per year, and no significant variation was found in the annual energy yield of microinverter-based PV systems. In Ref. [11], a sliding mode control (SMC) and reinforcement learning (RL) for a three-phase grid-integrated PV system are presented. The RL-based maximum power point tracking (MPPT) algorithm was implemented and SMC technique was used for reference current generation. Comparison between SMC-RL MPPT and fuzzy logic-based SMC was done. It was concluded that the SMC-RL technique performed better in terms of extraction and control of maximum power as compared to the fuzzy-based SMC method.

4.3   EXTREME LEARNING MACHINE (ELM)

In ELM, the weights and biases are selected in a random manner, and using predictors leads to the best solution. In ELM, the architecture is determined by an automatic mechanism in which a mathematical model is used to attributes of data using randomly allotted weights and biases. ELM uses feed-forward and backpropagation ANN, which improves the rate of convergence, generalization, tuning, and best fitting of data. In Ref. [12], ELM is employed to estimate worldwide sun radiation. In the testing phase, estimations made by ELM were compared with multiple linear regressions (MLR) and other models. Using ELM, the mean error and root mean square errors were found to be lower than that of MLR and other methods. ELM finds its application in power converters based PV systems where the objective is to improve the frequency stability of grid. In Ref. [13], a virtual inertia-based ML method is proposed for grid integration of PV system with improved frequency stability. Reinforcement learning (RL) is one of the popular forms of ML. Its background lies in learning mechanism of alive entities. RL performs actions on its environment and manipulates for the maximization of the received reward. The controller with ML technique holds advantages over the PI controller, such as ability to self-learn, capability to resolve the stability of frequency, quicker dynamic response, and adaptive control. Simulation studies showed that an ML-based controller was able to reduce error in steady state by 27% while maximum deviation of frequency with respect to nominal frequency was limited by 0.1 Hz. In Ref. [14], an algorithm that was a combination of RL with sleep schedule for coverage is applied. The proposed algorithm works in two stages. In the first stage, a precedence operator is assigned in a group and the formation of nodes is completed. In the next stage, a learning algorithm is generalized to a multi-sensor learning group and nodes are directed to work in a collaborative manner while adapting to the changing environment. The algorithm completes the learning of the entire team by changing the role of the active nodes while placing others in the sleep mode. The experimental results of the proposed algorithm on wireless sensors-based PV system suggests that the balance of energy consumption is maintained among the nodes, which results in increased life of the network along with desired coverage.

The output power of PV systems is generally dependent on the weather and climatic conditions. In Ref. [15], SVM learning with satellite technology was employed for predicting the output power of PV system. This prediction model was able to predict: (a) availability and quantity of clouds and (b) irradiance. Satellite pictures taken over the past four years were fed to configure the output and input data sets for SVM learning. The proposed model showed not only excellent accuracy in prediction as compared to conventional ANN models but also that prediction data can be used in grid-connected operation as well as in management of energy in the grid.

Identification of the type of fault and its location by using conventional methods is sometimes difficult in PV systems. This is due to variation in atmospheric conditions i.e. irradiance, shading, temperature, etc. In Ref. [16], a semi-supervised learning scheme was introduced for the detection, location finding, and classification of the fault, along with corrective measures taken to resolve the fault. The results exhibited that identification and correction of all the learned and unlearned faults were accomplished by the proposed method when PV arrays with prior experience were considered. The variations in output voltage were also minimized in the fault condition.

Nowadays, there is a rise in the demand for clean energy. The charging of electric vehicles (EV) is done using charging stations which are fed by PV systems. There is a requirement for the management of energy that can optimize the cost of operation and performance of energy storage systems. In this context, a new deep RL method that can handle time-varying data such as the charging status of the battery and data related to vehicle is proposed in Ref. [17]. This method provides the following advantages:

- Computation of the solutions of scheduling in EV charging stations
- Handling of time-varying data
- Ability to achieve desired performance
- Reduction in operating cost of charging stations

In Ref. [18], an ML-based electroluminescence (EL) imaging technique is proposed for PV systems. Using features of EL, prediction models of power and resistance were built for a PV system. The key benefit of these models is the requirement of EL imaging with PV module characterization and V–I curve only for the fast estimation of degraded PV modules. This technique offers other advantages such as more accurate prediction of efficiency, fast response, capability to electrical properties, and management and operation of PV systems. The developed models were found to be accurate in estimating the change in a series resistor, the degradation in the performance of a PV module, and the power of modules of different brands.

4.4   NEURAL NETWORK AI TECHNIQUES

In Ref. [19], a method of forecasting is proposed for the analysis of data obtained from a PV array. This method incorporates long-short term memory (LSTM) and CNN techniques together. This method can even provide the forecast for PV farms for the following day on whether some sensors are not in working condition or are not installed. In this methodology, selection of sufficient layers was done for the regressing the upcoming day's solar power. This was done on the basis of the previous day's data, i.e. values of irradiance and the panel's temperature taken every 10 minutes. The proposed method utilized the approximate data gathered from the meteorological centre and improved the performance of the PV system. In Ref. [20], a deep NN model is presented for accurate forecasting of a PV system's output power. It can produce a one-day forecasting of output power on the basis of input data such as irradiance, temperature, and past values of output power of PV systems. The simulation results showed that the proposed scheme was able to outperform other models in terms of accurate forecasting when subjected to highly irregular and unstable input data.

In Ref. [21], a conventional NN method is employed for pre-processing of EL images and dividing them into cells that serve as input data for ML algorithms. The pre-process of raw EL image data requires correction in lens distortion, filter and threshold operation, regression fitting, and transformation. Conventional NN performed better in terms of classification of PV cells as compared to random forest technique and SVM. In Ref. [22], an ANN method is proposed for predicting one day in advance irradiance curve under different meteorological conditions. This method improves the unprocessed forecast of numerical weather prediction by merging it with statistics-based learning schemes. The method was able to reduce the uncertainty in electrical power generation of PV systems. In Ref. [23], a forecasting method is presented for sun irradiance that overcame the limits of the LSTM method. This method encodes the time-series information into image form by combining LSTM with Gramian angular field. The advantage of this method is that it requires a small data set but provides excellent forecasting of irradiance.

In Ref. [24], a method for computation of global horizontal irradiance (GHI) has been proposed. Conventional methods for computing GHI involve costly equipment, workstations, and satellite-based models. Compared with traditional methods, this method is more accurate and convenient for prediction of GHI. In this work, CNN was applied on regression of images and detection and elimination of anomaly data was done using Gaussian and Bayesian models. The processed input having 3-month images and GHI data were fed into the proposed approach. Statistical analysis showed that the CNN-based method was more accurate, swift, inexpensive, and convenient for monitoring of big farms as compared to traditional irradiance measurement approaches.

The operational life of PV system can be increased if the diagnosis of fault is made properly. There are several conditions encountered such as shading, impedance of large value, mismatch of location, MPP tracking, and adverse weather that make the detection of faults difficult. In Ref. [25], a deep two-dimensional CNN is proposed for the extraction of features from 2-D graphs obtained from data of PV systems. The presented methodology achieved accuracy of approx. 74% for the detection of fault and approx. 70% for a noisy scenario. It is important to forecast a day advance for PV plants as it is desired for economic dispatch of power, energy management, and commitment of units. In Ref. [26], a similarity approach-based forecasting method is proposed for the prediction of PV output power in high resolution using low-resolution variables of weather. Their results demonstrated that the proposed methodology provided greater accuracy as compared to other forecasting models.

In Ref. [27], a method is proposed for short period scheduling of line maintenance in PV systems connected to a distribution network. It is well known that PV output power shows intermittent behaviour due to the involvement of random or fuzzy input variables that makes its prediction difficult. In the proposed strategy, initially, atmospheric data like irradiance and the amount of cloud cover are gathered. In the next step, uncertainties and scheduling of the distribution network are modelled with the help of a fuzzy program. It follows a pessimistic approach and optimizes the unenthusiastic values from the perspective of reliable and economic operation when subjected to probability constraints. In the last stage, a hybrid AI method was applied for finding the solution of the model. IEEE 33-Bus system was chosen for simulations and experiments. The outcomes of the proposed work validate the efficacy of the presented line maintenance for short-term scheduling of distribution network when subjected to uncertainties of fuzziness and randomness.

Adaptive Neuro-fuzzy interference system (ANFIS) is another powerful AI tool used for fast and dynamic MPPT control. In Ref. [28], ANFIS is applied for improving the accuracy and response of MPPT in standalone PV systems. The parameters controlled using the proposed scheme were injected power, severe value of voltage, frequency, and current. In Ref. [29], the role of internet of things (IoT)-based wireless sensor networks (WSN) in MPPT of PV arrays was discussed. The use of WSN in PV systems can reduce the requirement for complex hardware significantly, which results in low cost. In Ref. [30], an adaptive structure model is developed for forecasting a day in advance accessible energy in PV systems. The developed framework used a data analysis method that combined statistical and AI techniques.

In Ref. [31], the combined use of IoT with ANN and ANFIS is proposed for the prediction of power generated from PV systems. In terms of performance, ANN-based IoT is superior to ANFIS. In Ref. [32], AI was used for development of smart energy systems in cities. These studies all concluded that AI can help in improving the output of PV systems and in increasing the productivity of labour, which can pave the way for smart cities. In Ref. [33], an ANN-based model was implemented for controlling and dispatching steam power generation using a solar field. In this study, an ANN model having two BPs with four hidden layer networks was developed. The model utilized the prior inputs obtained from the steady-state heat transfer model. The first hidden 4 layers model the data on the solar field's temperature and pressure, while the second hidden layers perform analysis on the solar field's outlet temperature. The findings of this work suggest that the error in the prediction of pressure and temperature were reduced as compared to that obtained from the BP model and RNN while accuracy in simulation increased.

Management of energy is an essential goal to achieve in the development of a smart grid. This is because of integration of intermittent RE resources into the grid. Hence, the use of AI in smart grid is proposed to accurately forecast the generation from RE systems, including PV plants. In Ref. [34], a hybrid adaptive learning model is proposed which could make accurate predictions about the intensity of solar irradiance on the basis of weather data. In this research, a BP neural network model incorporating genetic algorithm was presented and applied on learning nonlinear associations in data. The presented methodology was able to detect the time-based, linearity/nonlinearity relationships in data, which resulted in enhancing predication ability. In simulation results, it was shown that presented methodology was superior to its counterparts in forecasting of solar intensity on a long-term or short-term basis. AI also plays an important role in the control and design of PV systems. In Ref. [35], recent AI methods used in the prediction of output power, diagnosis of faults, and controlling and calculation of optimal size in PV systems are reviewed. In the review, it was found that AI techniques such as artificial immune system (AIS), bee colony, and artificial fish swarm algorithm (AFSA) were more accurate in terms of identifying simulated annealing (SA) and GA. In a survey, NN was found to be superior in terms of accurate forecasting for MPPT and sizing PV systems and also in the diagnosis of fault.

In recent years, there has been bulk deployment of smart meters at the consumer end. The information gathered from these smart meters can be utilized for the forecasting of load demand. It can also improve the pattern of energy consumption adopted by consumers by using energy management methods. However, along with the many merits of smart meters, there are several challenges reported about this scheme. Data obtained through smart meter are volatile, variable, complex, and of large size. Management of this type of data is quite difficult. In Ref. [36], a new methodology is proposed based on clustering of data obtained through smart meter. This approach was presented for the fulfilment of the following objectives:

1. To fine detailed profile of load
2. To reduce complexity in load profile
3. To forecast load demand accurately
4. For optimum management of energy

In the presented concept of clustering, the data size of the smart meter was reduced. The simulation results validate that forecast of load was improved significantly using the clustering algorithm. In addition to this advantage, it also lightened the burden of mathematical calculation. It was also shown that the management of energy was improved as the proposed scheme resulted in larger savings of cost.

4.5   METAHEURISTICS-BASED AI TECHNIQUES

Metaheuristics techniques are inspired by natural or biological processes and play a major role in solving many engineering problems. Some of the contributions of metaheuristics are being discussed in this section. In Ref. [37], the problem of identification of PV model parameters was solved. This problem was solved by using a learning search algorithm (LSA). Salient attributes of the proposed LSA are as follows:

• As the iteration passes, rate of self-adjustment changes
• On the basis of the current worst and best solution, learning patterns get modified
• In the final stage, perturbation was allowed to ensure achieving of global optimum

The capability of the proposed LSA was compared to single-/double-diode PV models and other algorithms in terms of rate of convergence and computation time. Practical results suggest that the proposed technique performed well in finding the optimized parameters for PV systems. There is an important issue in PV systems to build a mathematical model suitable for optimized performance and reliable operation. In Ref. [38], a Tree Growth Algorithm (TGA) was implemented for finding unknown parameters of a PV model. TGA was able to find the precise values of parameters very quickly for the developed PV model. In TGA, there are two steps, namely exploration and exploitation. In first step, there is a competition among several trees to absorb the maximum sunlight for preparing their food. Initially, a population of trees was generated randomly; then, the fitness corresponding to every tree was determined. In the next step, some of the trees move toward the tree having the best fitness to receive abundant light. The trees having the worst fitness are removed and new trees take their places. As time passes, there will be movement of trees in the direction in which better food exists. In the next stage, most of the trees move towards best solutions and will be near to a global solution. There is a balance to be maintained between the stages of exploration and exploitation, and this is done by selection of proper parameters. The experimental values of PV systems' parameters obtained by TGA were found comparable to the datasheet provided by the manufacturer of the PV arrays.

There is an issue of finding a global MPP for the operation of PV systems under partial shading situations where PV characteristics have multiple optima points. Metaheuristics-based AI can conveniently solve this problem. In Ref. [39], a modified bat algorithm (BA) was proposed to achieve MPP quickly and efficiently. The study focused on two objectives:

- To reduce the time of MPP tracking for the case of sudden change in irradiance
- To converge at global MPP with a high level of confidence

In order to meet these two objectives, a combination of BA and cuckoo search (CS) algorithm was proposed. Results showed that proposed methodology was able to achieve global MPP accurately with higher rate of convergence as compared to BA and PSO. Metaheuristics-based AI also find their application in grid-integrated microgrid solutions. In Ref. [40], BA with a rule-based concept was proposed for the management of energy in the microgrid application. The microgrid under study comprised distributed units of PV modules, fuel cells, battery banks, and a microturbine. The proper functioning of the microgrid requires a day in advance prediction of not only intermittent data of PV modules and fuel cells but also modelling of the probabilistic nature of load. The prediction of PV data and load was done by this new BA algorithm. The forecasted data were fed to an energy management system on an hourly basis. It was shown that optimum management of the state of charge and reactive/active power were achieved.

In Ref. [41], optimization-based AI was used to control the microgrid based on a combination of PMDC generator, battery bank, and solar thermal power. The controllers adopted for the microgrid were required to control bi-directional charge and buck-boost control in converter. For this purpose, separate PI and PID controllers were used. The tuning of gain parameters of controllers was done by metaheuristics algorithms such as grey wolf (GW), mine blast, and particle swarm. In this work, simulation results demonstrated the excellent performance of GW optimization from the point of view of improving the rate of convergence, DC link voltage, state of charge in energy storage, and efficiency of converters, as compared to PSO and mine blast algorithm. In Ref. [42], an AI scheme based on marine predators (MP) was implemented for reconfiguration of PV modules. The proposed methodology was aimed at the mitigation of the effect of partial shading and at the prevention of hotspots forming in panels while achieving MPP. The proposed AI methods optimized the fitness function and factors such as losses due to mismatching of cells, fill factor, power loss, and improvement in power. The proposed algorithm showed significant dispersion of shade, resulting in the reduction of the count of peaks characteristic of PV power.

In Ref. [43], use of optimization technique-based AI is proposed for solving problems of predicting the characteristics of PV. The power output of the PV array fluctuates randomly due to variation in data obtained from the meteorological centre. This adversely affects the reliable and stable operation of the electrical grid. In order to overcome this issue, a hybrid algorithm based on the combination of ant-lion optimization (ALO) and a prediction method called random forest (RF) was proposed in the presented work. ALO optimized the parameters of the RF model, which led to reduction in computation time and improved accuracy. The proposed model was compared against the other RF models. It was found that accuracy in the prediction of performance of the PV system was improved significantly. In Ref. [44], a clustering methodology is presented based on expansion and erosion for finding faults in the PV system. In Ref. [45], a review of an AI-based forecasting method was done for accurate prediction of solar irradiance. The review also focused on key issues and future prospects for AI techniques in grid-integrated PV systems.

4.6 CONCLUSIONS

In this chapter, the role of various AI techniques in industrial and residential applications of PV systems has been discussed. The contributions of above discussed AI techniques is summarized in Table 4.1. It can be concluded from the tables that AI techniques played a crucial role in solving various issues of PV systems, and they have replaced the conventional analysis tools. In the future, AI techniques can play a major role in smart grid.

TABLE 4.1

Contribution of AI Techniques in PV Systems

Refs.	AI Technique Used	Problem Solved	Advantages
[5]	Ensemble-learning	• Efficiency of organic dye-sensitized solar cells	• Excellent generalization and accuracy • Superior to SVM and single base learner method
[6]	ANN Ensemble	• Forecasting for grid-tied PV system (231 kW)	• Accurate short-term prediction • Effective scheduling, reliable operation
[7]	Optimized adaptive residual ensemble model	• Forecast of power generation from PV system	• Accurate forecasting of solar power • Better prediction results as compared to ANN, SVM, and regression trees method

(Continued)

TABLE 4.1 (*Continued*)

Contribution of AI Techniques in PV Systems

Refs.	AI Technique Used	Problem Solved	Advantages
[9]	Deep learning (CNN, Auto encoder, BM)	• Investigation of energy policy • Review of DL techniques used in PV systems	• Analysis and processing of large data • Minimization of reconstruction error • Suitable for prediction of PV parameters
[10]	ML	• Forecast of PV generation by long-term analysis for partial shading conditions	• Long-term data analysis • No significant variation was found in annual energy yield of microinverter-based PV systems
[11]	RL-SMC	• MPPT in 3-phase grid-connected PV system	• Performed better in terms of extraction and control of MPP as compared to fuzzy-SMC
[12]	ELM	• Estimation of global solar radiation	• Reduced mean error and root mean square errors as compared to the regression method
[13]	RL	• Improvement of frequency stability for grid-tied PV system	• Faster dynamic response of controller • Reduced deviation in frequency at steady state, controller's ability to self-learn, and adaptive control
[16]	Semi supervised learning	• Fault classification and detection in PV systems	• Identification and correction of all the learned and unlearned faults • Fluctuation in voltage was minimized during fault.
[17]	deep RL	• Control of time-varying data related to PV-based EV	• Ease of computation in scheduling of EV charging stations and easy handling of data • Low operational cost of charging centres
[18]	EL imaging based ML	• Characterization of PV modules along with estimation of degraded PV modules	• Accurate prediction of efficiency, fast response, good estimation of series resistor • Estimation of degradation in PV module
[19]	LSTM based CNN	• Next-day forecast for the PV farms where sensors are not in working condition	• Accurate forecasting for PV systems when sensors were damaged • Improved the performance of PV system

(*Continued*)

TABLE 4.1 (*Continued*)

Contribution of AI Techniques in PV Systems

Refs.	AI Technique Used	Problem Solved	Advantages
[20]	Deep NN	• A day ahead forecasting of output power in PV systems based on input data	• Require data of irradiance, temperature, and past output power for prediction • Accurate forecast for highly irregular and unstable input data
[21]	Conventional NN	• Pre-process of raw EL image data for classification of PV cells.	• Performed better in terms of classification of PV cells as compared to RF technique and SVM
[22]	ANN	A day in advance prediction of solar irradiance curve	• Accurate prediction of irradiance using unprocessed forecast of weather prediction • Reduction in uncertainty of output power
[23]	LSTM	Forecasting for irradiance of Sun using a small data set	• Small data set required for forecasting • Excellent forecast of irradiance
[24]	CNN	Prediction of GHI for monitoring large PV farms	• Accurate, fast, inexpensive, and convenient for monitoring of big PV farms
[25]	2-D CNN	Feature extraction from graphs obtained from PV array	• Economic dispatch, energy management, high accuracy in detection of fault
[27]	Fuzzy logic AI	Short-period line maintenance scheduling in PV systems	• Short-term scheduling achieved in distribution network with uncertainties of fuzziness and randomness.
[28]	ANFIS	Improved accuracy and response of MPPT in standalone PV systems	• Injected power, severe value of voltage, frequency, and current were controlled using ANFIS
[29]	IoT	IoT-based WSN in MPPT of PV arrays	• The need for complex requirement of hardware was reduced, which results in low cost
[30]	Adaptive AI	A day ahead accessible energy forecast in PV systems	• Developed framework used data analysis method with statistical and AI techniques
[31]	IoT with ANN and ANFIS	Prediction of power generated from PV systems	• Proposed technique was superior to ANFIS
[33]	ANN	Control and dispatch steam power using solar fields	• Error in prediction of pressure and temperature were reduced

(*Continued*)

TABLE 4.1 (*Continued*)

Contribution of AI Techniques in PV Systems

Refs.	AI Technique Used	Problem Solved	Advantages
[34]	GA based BP NN	Accurate prediction of intensity of solar irradiance	• Enhanced predication ability • Excellent long-/short-term forecasting
[37]	LSA	Identification of PV model parameters	• Optimized parameters for PV systems were found with a high rate of convergence
[38]	TGA	Finding unknown parameters of the PV model	• Parameters obtained were comparable to that in the datasheet provided by the manufacturer
[39]	BA-cuckoo search algorithm	Finding global MPP for PV systems under shading cases	• Reduced time in MPP tracking for sudden variation in irradiance • High probability of finding MPP
[40]	Rule base BA	Management of energy in microgrid	• Accurate prediction of PV data and load • Optimum management of state of charge and reactive/active power
[41]	Grey wolf (GW) based AI	Control of Microgrid	• Tuning of gain parameters of controllers • Improved voltage stability and SOC in energy storage systems
[42]	Marine predators based AI	Reconfiguration of PV modules	• Mitigation of adverse effects of partial shading while achieving MPPT • Prevention of hotspot formation
[43]	ALO-RF based AI	Predict PV characteristics based on meteorological data for stable and reliable grid	• Optimized the parameters of RF model • Less computation time and improved accuracy

REFERENCES

1. R. Ghannam, P. V. Klaine and M. Imran, "Artificial Intelligence for Photovoltaic Systems," In: Precup, R. E., Kamal, T., and Zulqadar Hassan, S. (eds.), *Solar Photovoltaic Power Plants.* Springer, Singapore, 2019.
2. R. Belu, "Artificial Intelligence Techniques for Solar Energy and Photovoltaic Applications," *Handbook of Research on Solar Energy Systems and Technologies*, vol. 3, pp. 376–436, 2012. Doi:10.4018/978-1-4666-1996-8.ch015.
3. K. AbdulMawjood, S. S. Refaat and W. G. Morsi, "Detection and Prediction of Faults in Photovoltaic Arrays: A Review," *2018 IEEE 12th International Conference on Compatibility, Power Electronics and Power Engineering (CPE-POWERENG 2018)*, Doha, 2018, pp. 1–8.

4. N. L. Georgijevic, M. V. Jankovic, S. Srdic and Z. Radakovic, "The Detection of Series Arc Fault in Photovoltaic Systems Based on the Arc Current Entropy," in *IEEE Transactions on Power Electronics*, vol. 31, no. 8, pp. 5917–5930, Aug. 2016.

5. H. Li et al., "Ensemble Learning for Overall Power Conversion Efficiency of the All-Organic Dye-Sensitized Solar Cells," in *IEEE Access*, vol. 6, pp. 34118–34126, 2018.

6. S. Al-Dahidi, O. Ayadi, M. Alrbai and J. Adeeb, "Ensemble Approach of Optimized Artificial Neural Networks for Solar Photovoltaic Power Prediction," in *IEEE Access*, vol. 7, pp. 81741–81758, 2019.

7. H. Su, T. Liu and H. Hong, "Adaptive Residual Compensation Ensemble Models for Improving Solar Energy Generation Forecasting," in *IEEE Transactions on Sustainable Energy*, vol. 11, no. 2, pp. 1103–1105, Apr. 2020.

8. M. Khodayar, S. Mohammadi, M. E. Khodayar, J. Wang and G. Liu, "Convolutional Graph Autoencoder: A Generative Deep Neural Network for Probabilistic Spatio-Temporal Solar Irradiance Forecasting," in *IEEE Transactions on Sustainable Energy*, vol. 11, no. 2, pp. 571–583, Apr. 2020.

9. S. Shamshirband, T. Rabczuk and K. Chau, "A Survey of Deep Learning Techniques: Application in Wind and Solar Energy Resources," in *IEEE Access*, vol. 7, pp. 164650–164666, 2019.

10. N. T. Le and W. Benjapolakul, "Comparative Electrical Energy Yield Performance of Micro-Inverter PV Systems Using a Machine Learning Approach Based on a Mixed-Effect Model of Real Datasets," in *IEEE Access*, vol. 7, pp. 175126–175134, 2019.

11. A. Bag, B. Subudhi and P. K. Ray, "A Combined Reinforcement Learning and Sliding Mode Control Scheme for Grid Integration of a PV System," in *CSEE Journal of Power and Energy Systems*, vol. 5, no. 4, pp. 498–506, Dec. 2019.

12. T. Hai et al., "Global Solar Radiation Estimation and Climatic Variability Analysis Using Extreme Learning Machine Based Predictive Model," in *IEEE Access*, vol. 8, pp. 12026–12042, 2020.

13. K. Y. Yap, C. R. Sarimuthu and J. M. Lim, "Grid Integration of Solar Photovoltaic System Using Machine Learning-Based Virtual Inertia Synthetization in Synchronverter," in *IEEE Access*, vol. 8, pp. 49961–49976, 2020.

14. H. Chen, X. Li and F. Zhao, "A Reinforcement Learning-Based Sleep Scheduling Algorithm for Desired Area Coverage in Solar-Powered Wireless Sensor Networks," in *IEEE Sensors Journal*, vol. 16, no. 8, pp. 2763–2774, Apr. 15, 2016.

15. H. S. Jang, K. Y. Bae, H. Park and D. K. Sung, "Solar Power Prediction Based on Satellite Images and Support Vector Machine," in *IEEE Transactions on Sustainable Energy*, vol. 7, no. 3, pp. 1255–1263, July 2016.

16. H. Momeni, N. Sadoogi, M. Farrokhifar and H. F. Gharibeh, "Fault Diagnosis in Photovoltaic Arrays Using GBSSL Method and Proposing a Fault Correction System," in *IEEE Transactions on Industrial Informatics*, vol. 16, no. 8, pp. 5300–5308, Aug. 2020.

17. M. Shin, D. Choi and J. Kim, "Cooperative Management for PV/ESS-Enabled Electric Vehicle Charging Stations: A Multiagent Deep Reinforcement Learning Approach," in *IEEE Transactions on Industrial Informatics*, vol. 16, no. 5, pp. 3493–3503, May 2020.

18. A. M. Karimi et al., "Generalized and Mechanistic PV Module Performance Prediction from Computer Vision and Machine Learning on Electroluminescence Images," in *IEEE Journal of Photovoltaics*, vol. 10, no. 3, pp. 878–887, May 2020.

19. W. Lee, K. Kim, J. Park, J. Kim and Y. Kim, "Forecasting Solar Power Using Long-Short Term Memory and Convolutional Neural Networks," in *IEEE Access*, vol. 6, pp. 73068–73080, 2018.

20. C. Huang and P. Kuo, "Multiple-Input Deep Convolutional Neural Network Model for Short-Term Photovoltaic Power Forecasting," in *IEEE Access*, vol. 7, pp. 74822–74834, 2019.

21. A. M. Karimi et al., "Automated Pipeline for Photovoltaic Module Electroluminescence Image Processing and Degradation Feature Classification," in *IEEE Journal of Photovoltaics*, vol. 9, no. 5, pp. 1324–1335, Sept. 2019.

22. L. C. Parra Raffán, A. Romero and M. Martinez, "Solar Energy Production Forecasting Through Artificial Neuronal Networks, Considering the Föhn, North and South Winds in San Juan, Argentina," in *The Journal of Engineering*, vol. 2019, no. 18, pp. 4824–4829, July 2019.

23. Y. Hong, J. J. F. Martinez and A. C. Fajardo, "Day-Ahead Solar Irradiation Forecasting Utilizing Gramian Angular Field and Convolutional Long Short-Term Memory," in *IEEE Access*, vol. 8, pp. 18741–18753, 2020.

24. H. Jiang, Y. Gu, Y. Xie, R. Yang and Y. Zhang, "Solar Irradiance Capturing in Cloudy Sky Days–A Convolutional Neural Network Based Image Regression Approach," in *IEEE Access*, vol. 8, pp. 22235–22248, 2020.

25. F. Aziz, A. Ul Haq, S. Ahmad, Y. Mahmoud, M. Jalal and U. Ali, "A Novel Convolutional Neural Network-Based Approach for Fault Classification in Photovoltaic Arrays," in *IEEE Access*, vol. 8, pp. 41889–41904, 2020.

26. H. Sangrody, N. Zhou and Z. Zhang, "Similarity-Based Models for Day-Ahead Solar PV Generation Forecasting," in *IEEE Access*, vol. 8, pp. 104469–104478, 2020.

27. Z. Bao, C. Gui and X. Guo, "Short-Term Line Maintenance Scheduling of Distribution Network with PV Penetration Considering Uncertainties," in *IEEE Access*, vol. 6, pp. 33621–33630, 2018.

28. H. Abu-Rub, A. Iqbal, S. Moin Ahmed, F. Z. Peng, Y. Li and G. Baoming, "Quasi-Z-Source Inverter-Based Photovoltaic Generation System with Maximum Power Tracking Control Using ANFIS," in *IEEE Transactions on Sustainable Energy*, vol. 4, no. 1, pp. 11–20, Jan. 2013.

29. A. Omairi, Z. H. Ismail, K. A. Danapalasingam and M. Ibrahim, "Power Harvesting in Wireless Sensor Networks and Its Adaptation with Maximum Power Point Tracking: Current Technology and Future Directions," in *IEEE Internet of Things Journal*, vol. 4, no. 6, pp. 2104–2115, Dec. 2017.

30. Y. S. Manjili, R. Vega and M. M. Jamshidi, "Data-Analytic-Based Adaptive Solar Energy Forecasting Framework," in *IEEE Systems Journal*, vol. 12, no. 1, pp. 285–296, Mar. 2018.

31. V. Puri et al., "A Hybrid Artificial Intelligence and Internet of Things Model for Generation of Renewable Resource of Energy," in *IEEE Access*, vol. 7, pp. 111181–111191, 2019.

32. A. C. Şerban and M. D. Lytras, "Artificial Intelligence for Smart Renewable Energy Sector in Europe—Smart Energy Infrastructures for Next Generation Smart Cities," in *IEEE Access*, vol. 8, pp. 77364–77377, 2020.

33. S. Guo, H. Pei, F. Wu, Y. He and D. Liu, "Modeling of Solar Field in Direct Steam Generation Parabolic Trough Based on Heat Transfer Mechanism and Artificial Neural Network," in *IEEE Access*, vol. 8, pp. 78565–78575, 2020.

34. Y. Wang, Y. Shen, S. Mao, G. Cao and R. M. Nelms, "Adaptive Learning Hybrid Model for Solar Intensity Forecasting," in *IEEE Transactions on Industrial Informatics*, vol. 14, no. 4, pp. 1635–1645, Apr. 2018.

35. Ayman Youssef, Mohammed El-Telbany and Abdelhalim Zekry, "The Role of Artificial Intelligence in Photovoltaic Systems Design and Control: A Review," *Renewable and Sustainable Energy Reviews*, vol. 78, 2017, pp. 72–79, 2017.

36. Z. A. Khan and D. Jayaweera, "Smart Meter Data Based Load Forecasting and Demand Side Management in Distribution Networks With Embedded PV Systems," in *IEEE Access*, vol. 8, pp. 2631–2644, 2020.

37. T. Huang, C. Zhang, H. Ouyang, G. Luo, S. Li and D. Zou, "Parameter Identification for Photovoltaic Models Using an Improved Learning Search Algorithm," in *IEEE Access*, vol. 8, pp. 116292–116309, 2020.

38. A. A. Z. Diab, H. M. Sultan, R. Aljendy, A. S. Al-Sumaiti, M. Shoyama and Z. M. Ali, "Tree Growth Based Optimization Algorithm for Parameter Extraction of Different Models of Photovoltaic Cells and Modules," in *IEEE Access*, vol. 8, pp. 119668–119687, 2020.
39. C. Y. Liao, R. K. Subroto, I. S. Millah, K. L. Lian and W. Huang, "An Improved Bat Algorithm for More Efficient and Faster Maximum Power Point Tracking for a Photovoltaic System Under Partial Shading Conditions," in *IEEE Access*, vol. 8, pp. 96378–96390, 2020.
40. M. Elgamal, N. Korovkin, A. Elmitwally, A. A. Menaem and Z. Chen, "A Framework for Profit Maximization in a Grid-Connected Microgrid with Hybrid Resources Using a Novel Rule Base-BAT Algorithm," in *IEEE Access*, vol. 8, pp. 71460–71474, 2020.
41. R. Kumar and N. Sinha, "Modeling and Control of Dish-Stirling Solar Thermal Integrated with PMDC Generator Optimized by Meta-Heuristic Approach," in *IEEE Access*, vol. 8, pp. 26343–26355, 2020.
42. D. Yousri, T. S. Babu, E. Beshr, M. B. Eteiba and D. Allam, "A Robust Strategy Based on Marine Predators Algorithm for Large Scale Photovoltaic Array Reconfiguration to Mitigate the Partial Shading Effect on the Performance of PV System," in *IEEE Access*, vol. 8, pp. 112407–112426, 2020.
43. I. A. Ibrahim, M. J. Hossain and B. C. Duck, "An Optimized Offline Random Forests-Based Model for Ultra-Short-Term Prediction of PV Characteristics," in *IEEE Transactions on Industrial Informatics*, vol. 16, no. 1, pp. 202–214, Jan. 2020.
44. L. Djilali, E. N. Sanchez, F. Ornelas-Tellez, A. Avalos and M. Belkheiri, "Improving Microgrid Low-Voltage Ride-Through Capacity Using Neural Control," in *IEEE Systems Journal*, vol. 14, no. 2, pp. 2825–2836, June 2020.
45. A. T. Eseye, M. Lehtonen, T. Tukia, S. Uimonen and R. J. Millar, "Adaptive Predictor Subset Selection Strategy for Enhanced Forecasting of Distributed PV Power Generation," in *IEEE Access*, vol. 7, pp. 90652–90665, 2019.

5 Thermodynamic Approaches and Techniques for Solar Energy Conversion

Ankit Dasgotra, Vishal Kumar Singh, and Sunil Kumar Tiwari
University of Petroleum & Energy Studies, Dehradun, India

Raj Kumar Mishra
IIT Roorkee, Roorkee, India

CONTENTS

5.1 INTRODUCTION

As an emerging thrust on the economy, developing energy sectors are playing an important role (Ozturk et al., 2010). Energy demand remains unpredictable and it widens the existence of demand vs supply gap (Z. Liu, 2015). The major concern worldwide remains the misuse of fossil fuels, which leads to global warming and atmospheric pollution (Barbier, 2010; Zaman et al., 2014).

These elements roused the global analysts to rally for a green energy innovation, i.e. renewable power sources (Kalogirou, 2013) from different inexhaustible wellsprings of energy, for example, solar energy (Al-Karaghouli et al., 2009), bioenergy (Z. Liu, 2015), geothermal vitality (Coskun et al., 2012), etc. Solar-oriented innovation has developed as a significant hotspot for the transformation of sun-based energy into thermal and electrical energy, and this provides an unlimited source of power and is also ecological (Philibert, 2011). It has been stated that the amount of sun rays falling on earth every hour can satisfy the world's energy needs for the entire year. However, the acknowledgement of sun-based energy as a substitute wellspring of energy is indicated as a direct result of its higher activity cost and lower effectiveness; thus, different explorations to exploit this was done (S. Liu et al., 2014) [10]. The solar collector is one of such gadget that changes sun-based energy into thermodynamic energy, utilizing a heat-trading fluid as the absorbed fluid (Minardi et al., 1975).

5.2 IMPORTANCE OF SOLAR ENERGY

Nowadays, renewable energies have achieved significant importance because of the fast consumption and exhaustion of fossil fuels, the increase of greenhouse gas emissions, and the weather variations. Currently, approx. 82% of the global basic energy necessity is obtained from coal and natural gas. (*Key World Energy Statistics 2020 – Analysis – IEA*, 2020). Almost 12% originates from biomass and 6% from hydroelectric sources. The greenhouse effect can be reduced as we start relying on renewable resources, and solar energy is one such effective way.

5.3 NATURE OF SOLAR ENERGY

Any type of energy mainly generated by the sun can be stated as solar energy. Nuclear fusion (hydrogen protons collide and their fusion generates helium atom) occurring in the sun helps to generate solar energy. This process emits tremendous energy as the fusion of 620 million metric tons of energy takes place every second inside the sun's core (Morse & Turgeon, 2012) (Kikuchi et al., 2012). Sun rays travel in the form of electromagnetic radiation. The electromagnetic radiation travels in the form of waves that carry energy, heat, and light, having different spectral frequency and wavelength. Figure 5.1 shows a representation of the basic nuclear fusion process in the sun, giving direct and zigzag pathways for energy transfer (Kikuchi et al., 2012).

The most high-recurrence waves produced by the sun are gamma beams, X-rays, and UV rays. The most unsafe ultraviolet rays is consumed by the Earth's environment. Less intense UV radiations travel through the atmosphere and can cause

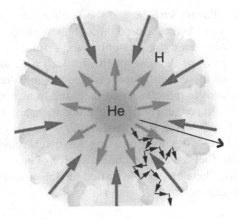

FIGURE 5.1 Nuclear fusion process in the sun (Kikuchi et al., 2012).

sunburn. The sun additionally produces infrared radiation, which has a lower recurrence. Most of the warmth from the sun shows up as infrared. Sandwiched among infrared and UV is the noticeable visible range, which contains all the hues we see on Earth. The color red has the longest frequency (nearest to infrared), and violet (nearest to UV) the shortest (Kikuchi et al., 2012; Morse & Turgeon, 2012).

Solar energy exists in the form of infrared rays, visible rays, and Ultra-Violet spectrum waves that reach our planet to warm it, thus making the existence of life possible on earth. Approx. 30% of energy is sent back to space, and the remaining portion is utilized on earth after it passes through the atmosphere (Morse & Turgeon, 2012). Figure 5.2 shows a complete representation of solar radiation utilized by earth.

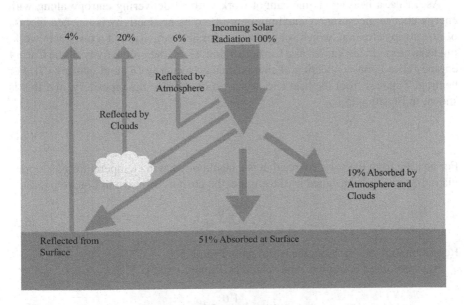

FIGURE 5.2 Utilization of incoming solar radiation (Morse & Turgeon, 2012).

Radiations heat the Earth's surface, and the surface emanates a portion of this back out as infrared waves. As they ascend through the atmosphere, they are blocked by ozone-depleting substances, for example, water fumes and carbon dioxide. Photosynthesis takes place as the absorption of solar rays and its conversion into nutrients take place. All fossil fuel on earth is also possible due to the existence of photosynthesis. This helps to preserve fossil fuels and transforms them into micro-organisms that under high pressure and temperature are again transformed to petroleum, natural gas and coal. Morse & Turgeon, 2012.

Humans evaluated processes for extracting fossil fuels and utilizing them for energy. However, fossil fuel is a non-renewable source of energy, as their formation takes millions of years. Therefore, people started relying on direct sources of renewable energy like solar energy.

5.4 THERMODYNAMIC APPROACH

Radiation is the main source of solar energy transferring heat (5800 Kelvin) to the earth. The second law of thermodynamics is involved in the principle of converting heat energy into other kinds of energy, i.e. mechanical, chemical, electrical, etc. Heating engines are mainly utilized for conversion as they absorb energy (in the form of current), i.e. I_{in} along with thermal heat T_S and an entropy current I_S. The engine needs to dispose the consumed entropy and dispose-off the entropy conceivably created during the time spent in transformation via providing the current $I_{S,R}$ to a temperature reservoir T. Along $I_{S,R}$, an energy current $I_{E,R}$ exists and goes to the heat reservoir as entropy can only exist in heat formation. Remaining energy existed in the form of current having zero entropy due to engine production can be stated as electrical energy ($I_{E,E}$) via solar cell or chemical energy ($I_{E,C}$) because of photosynthesis.

As a rule, a heating engine cannot work without delivering entropy along with energy change measures. A maximum cut-off to its productivity is reached in case of no entropy creation, which is known as reversible activity, and a reversible working heating engine is known as a Carnot engine. Here, the productivity of the Carnot engine relies upon connection of the existence of energy (E) and entropy (S). For heating, T moves from the heat engine via conduction to the repository, and this is shown in Equation (5.1):

$$I_{E,R} = T I_{S,R} \tag{5.1}$$

For heating from the solar source that is a blackbody having temperature T_S, shipped via radiations and consumed by the engine, the current is shown in Equation (5.2):

$$I_{E,R} = A \frac{\theta}{\pi} \sigma \tag{5.2}$$

Representation of a heat engine is shown in Figure 5.3.

Also, absorbed entropy is represented by Equation (5.3):

$$I_S = A \frac{4}{3} \frac{\theta}{\pi} \sigma T^3 \tag{5.3}$$

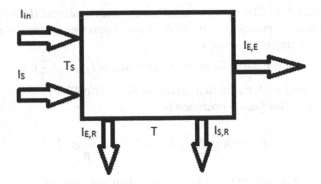

FIGURE 5.3   Representation of a heat engine (P. Würfel, 2002; Würfel, 1988).

where θ = solid angle specified by the sun = 6.8×10^{-5} for non-concentrating solar light and it is equal to π as maximum focus onto a planar black absorbing surface.

In addition, σ = Stefan–Boltzmann constant = 5.67×10^{-8} W/ $(m^2 K^4)$.

The connection between energy and entropy is not the same as in Equation (5.1) because of inescapable entropy production when radiation is discharged into free space. Here, entropy vanishes when a similar radiation is produced that is additionally retained, just like the body is in thermal harmony with its environmental factors (P Würfel, 2002; Würfel, 1988).

The reversible process of the heat engine is represented in Equation (5.4):

$$I_S = I_{S,R} \tag{5.4}$$

As an output of heating engine, electric energy formation exists, and this is shown in Equation (5.5):

$$I_{E,E} = I_E - I_{E,R} = I_E - T I_S = I_E \left(1 - \frac{4}{3} \frac{T}{T_S} \right) \tag{5.5}$$

Also, the efficiency of the reversible heating engine is given in Equation (5.6):

$$\eta = \frac{I_{E,E}}{I_E} = 1 - \frac{4}{3} \frac{T}{T_S} = 1 - \frac{4}{3} \frac{300\ K}{5800\ K} = 0.931034 \tag{5.6}$$

Interestingly, this cut-off effectiveness does not rely upon whether the radiations from the sun is concentrated, demonstrating that the higher caliber of sun-based vitality is available as of now in non-concentrating sunlight. It is known that a body that absorbs radiation should likewise produce radiations based on its temperature. Also, Kirchhoff's law expresses that the emissivity of an object for radiations of a specific frequency is equivalent to its absorptivity at a similar frequency. Landsberg (De Vos et al., 1993) looked at that as an engine at T_a transmitting radiations, and this must be remembered for the vitality and the entropy balance that includes an emitter current $I_{E,o}$ and a produced entropy-based current $I_{S,o}$, both at temperature T_a (as shown in Figure 5.3). The occurrence of entropy-based current (Equation 5.3) that the engine retains is presently halfway discharged into free space and is somewhat moved to the heating reservoir at T.

The released bit is passing from the A into π and is $I_{S,O} = A\dfrac{4}{3}\sigma T_a^3$.

It is stated along with an emitted current of $I_{E,O} = A\sigma T_a^4$.

Energy balance for the rev operation is

$$I_{S,O} = A\frac{4}{3}\sigma T_a^3 + I_{S,R} = I_S = A\frac{4}{3}\frac{\theta}{\pi}\sigma T_S^3$$

Entropy and Energy current to the heating reservoir are given as

$$I_{S,R} = A\frac{4}{3}\sigma\left(\frac{\theta}{\pi}T_S^3 - T_a^3\right) \text{ And } I_{E,R} = TI_{S,R}$$

Now, the energy balance is shown in Equations (5.7) and (5.8):

$$I_{E,E} = I_E - I_{E,O} - I_{E,R} = A\frac{\theta}{\pi}\sigma T_S^4\left(1 - \frac{\pi}{\theta}\frac{T_a^4}{T_S^4} - \frac{4}{3}\frac{T}{T_S}\left[1 - \frac{\pi}{\theta}\frac{T_a^3}{T_S^3}\right]\right) \quad (5.7)$$

and

$$\eta_L = 1 - \frac{\pi}{\theta}\frac{T_a^4}{T_S^4} - \frac{4}{3}\frac{T}{T_S}\left[1 - \frac{\pi}{\theta}\frac{T_a^3}{T_S^3}\right] \quad (5.8)$$

For $\theta = \pi$, it achieves the Landsberg efficiency [17]. Its maximum efficiency is 0.931036 9 (η_L) is attained if the absorbing body exists at room conditions, i.e. $T_a = T$.

Generation of entropy takes place if a certain amount of external current, such as a particle or charge, is moving around a variance of quantities like temperature and chemical or electrical potentials. There also exists a situation wherein heating from the sun could be absorbing but not be associated with the production of entropy; there also exist another condition whereby the absorbing body is in equilibrium position along radiation from the lunar source.

5.5 THERMAL EQUILIBRIUM

Equilibrium of temperature (i.e. thermal equilibrium) exists as a dual system having the same temperatures, but with diverse pressures and diverse chemical potentiality. Similarly, the equilibrium of pressure needs equivalent pressures. Thus, it is necessary to mention these types of equilibria to distinguish their assigning. Thermodynamic equilibrium exists as a sum of temperature, pressure, and chemical equilibrium; but its existence does not appear in real scenarios. Two systems/objects are in thermal equilibrium if they have similar temperature and there is heat exchange between the systems as a reversible process, as shown in Figure 5.4.

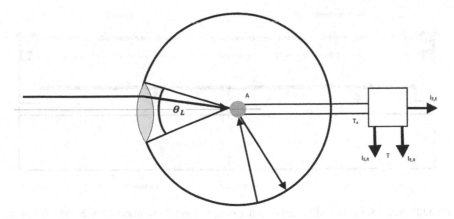

FIGURE 5.4 Heat engine consists of a blackbody absorber (*A*) that gets the sun via lens or itself. Its deliveries heat T_a to a Carnot engine (P. Würfel, 2002; Würfel, 1988).

5.5.1 SOLAR COLLECTOR (TYPES AND APPLICATIONS)

A solar collector is defined as an operator that collects radiation from the sun. These are used for active heating for domestic usage (Nkwetta et al., 2010). In other words, this device transforms solar energy into heat by means of a solar collector.

The usage of such collectors acts as a substitute for the conventional household water heater with a heater, and is also economical in the long run. Apart from domestic use, various solar collectors can be united in series to produce significant amounts of energy.

Solar collectors are generally categorized into two groups:

a. non-concentrating
b. concentrating.

Non-concentrating: The gatherer area is similar to that of the absorbing space (Weiss & Rommel, 2008a). Flat plate collector (FPC) and evacuated tube collector (ETC) belong to the non-concentrating category. These are specially considered for industry-based heating claims that require energy output at temperature ranges from 60°C to 2500°C. They are structurally modest and need lesser care.

5.5.2 FLAT PLATE SOLAR WATER COLLECTOR

The illustration of an FPC is given in Figure 5.5. It contains (a) an absorber, (b) cover, (c) transporting fluid, (d) housing, and (e) insulation. Better conductivity is needed for transmission of the heat from the absorbing sheet to the absorbing pipe. Usually, mineral wool is utilized as insulation, reducing the heat loss at the absorbing surface (Weiss & Rommel, 2008b). These can be attained via:

i. Multiple glazing with anti-reflective glass,
ii. Filling a hermetically sealed flat plate collector with a noble gas, or
iii. Evacuating a hermetically sealed flat plate collector (Seraphin, 1976).

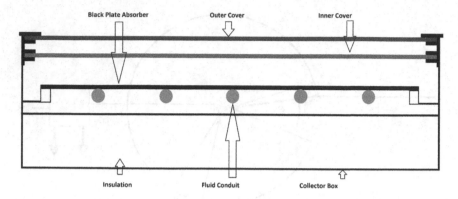

FIGURE 5.5 Schematic diagram of a flat-plate collector (Kennedy, 2002; Pekruhn et al., 1985; Redaelli, 1976).

These modified FPCs can be utilized for solar-industrial-process-heating (SHIP) claims (Kennedy, 2002; Pekruhn et al., 1985; Redaelli, 1976).

5.5.3 Evacuated Tube Collector

Figure 5.6 shows the layout of an ETC. Each evacuated cylinder has dual glass tubes produced using powerful borosilicate glass with higher chemicals and warm stun obstruction. The external side of the cylinder is straight forward, allowing light beams to pass through with negligible reflection. The outside of the inner cylinder is covered with a faltered sun powered particular covering that is characterized by high sun-based absorptance and low warmth emittance. The top finishes of the two cylinders are combined and the air caught in the annular space between the two layers is emptied to limit conductive and convective heating losses. The top finish of these parallel cylinders is fitted to the inward stockpiling tank. In making a vacuum, a barium getter is embedded into the base of the external glass tube. The inside

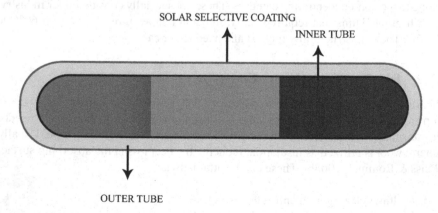

FIGURE 5.6 Schematic illustration of an ETC (Kennedy, 2002; Pekruhn et al., 1985; Redaelli, 1976).

glass tube is then embedded into the outside cylinder with the getter focusing on the inward glass tube. The glass tubes are raised to a higher temperature and a vacuum is produced. The two glass tubes are then joined at the open end. The barium getter likewise fills extra needs. At the point when the glass tubes are warmed, before the finishes are melded, the barium getter becomes very hot and an unadulterated layer of barium is deposited at the base of the cylinder that will appear as though a chrome plate is placed within the external glass tube.

The benefits of an ETC in comparison to an FPC is that the regular outline of the rotund evacuated tube remains perpendicular to the solar radiations. Thus, energy is nearly constant.

In the concentrating type, different kinds of mirrors, reflectors, or concentrators are employed, concentrating the sun-based energy and providing high temperatures (i.e., 400°C–1000°C), than that achievable in non-concentrating types.

There are four types:

a. Compound parabolic concentrator

CPC collector needs a larger quantity of sunlit and requires lesser precise tracing when compared to PTC. Hence, it fills the gap among the lower-temperature FPCs ($T < 800°C$) and the higher-temperature concentrating concentrators ($T > 4000°C$). For sun-based temperature production, it is needed to create vitality at a temperature higher than that of ETCs (non-concentrated) and CPC gatherers. A concentrating gatherer can be utilized for higher-temperature claims, for example, to produce steam. Sun-based force frameworks, otherwise called concentrated sunlight-based force frameworks, use concentrated sun-oriented radiation as a higher-temperature vitality source to create power utilizing a warm course. Direct sun-oriented radiation can be focused and gathered by the scope of CSP innovations to give a scope of medium to high warmth prerequisites, as shown in Figure 5.7.

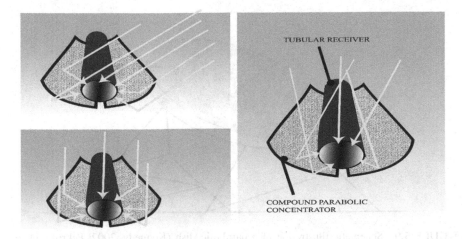

FIGURE 5.7 Schematic illustration of a compound parabolic concentrator collector (Kennedy, 2002; Pekruhn et al., 1985; Redaelli, 1976).

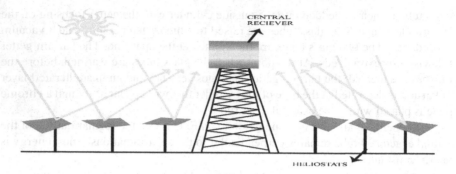

FIGURE 5.8 Schematic diagram of central receiver (or) solar tower (Kennedy, 2002; Pekruhn et al., 1985; Redaelli, 1976).

  b. Central receiver/Solar tower

     A roundabout exhibit of heliostats (enormous separately following mirrors) is utilized to focus daylight onto a focal beneficiary fixed on the head of a pinnacle. A warmth move medium in the focal recipient gets the higher concentrated ray replicated via heliostats and change it into warm vitality, which is then utilized for the age of superheated steam for turbine activity (Figure 5.8).

  c. Parabolic dish collector

     An explanatory dish-molded reflector is utilized to focus daylight onto a beneficiary set at the point of convergence of the dish. The concentrating beam rays are absorbed into the beneficiary to heat up a liquid or gas (air) to roughly 750°C. This liquid or gas is then utilized to deliver power through a little cylinder or a microturbine, fixed to the beneficiary, as shown in Figure 5.9.

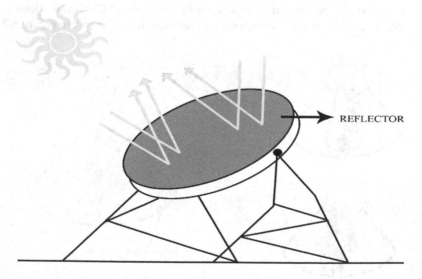

FIGURE 5.9 Schematic illustration of a parabolic dish (Kennedy, 2002; Pekruhn et al., 1985; Redaelli, 1976).

d. Parabolic trough collector

These collectors utilize trough-shaped mirror reflector to focus solar light on the receiver tube with the help of which a fluid is heated up to approximately 400°C, which is then utilized to generate super-heated steam.

5.6 SOLAR PV SYSTEMS AND APPLICATIONS

The PV module is a rapid technological innovation that will reduce the PV feed-in tariff in future. Photovoltaics is a technology that converts solar energy into electrical energy. International experience has been gained over many years in manufacturing this, resulting in updated Photovoltaic module effectiveness, price reduction, and increase in productivity (Gong & Kulkarni, 2005). The electricity generated is self-consumed up to a certain percentage (Moshövel et al., 2015). If an exceptionally high amount of sun-based electrical energy is generated, it is stored in a battery and could be used in future when it is not covering the electrical capacity.

The photovoltaic systems are divided into four major applications.

5.6.1 Domestic Off-Grid Photovoltaic System

These photovoltaic systems can be used to provide electricity to the houses and villages that are remotely located and not linked to the nationwide network system. These schemes usually deliver power to the equipment that run on low power loads such as lighting and refrigerators. Thousands of PV systems are being installed worldwide to meet the electricity demand of the off-grid community. It is an extensive alternative for extending the electricity distribution in areas that are difficult to access (Mills, 2000; Nkwetta et al., 2010). When the Photovoltaic pattern is active for local utilization of the Photovoltaic units are mounted on rooftop, which reduces the land requirements.

5.6.2 Non-Domestic Off-Grid Photovoltaic System

This is the most appropriate application, especially when electricity is in high demand and PV is cost-competitive in comparison with other electricity-generating sources. These PV systems are in high demand in industries, airports, and other commercial sectors. They provide power for use in low-maintenance systems such as telecommunication, water pumping, and navigation aids.

5.6.3 Grid Connecting Photovoltaic Systems

These photovoltaic systems are connected to the grids and supply power to the building or any other power requirement equipment. This system is integrated with industrial buildings and residential houses. There is no requirement for battery storage as the system is connected to the grid and provide output varying between 1 and 100 kW. When the onsite generation exceeds the demand, the electricity is fed back to the grid (Zahedi, 1997, 2004).

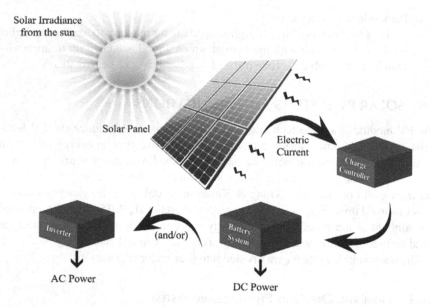

Solar Irradiance from the sun

Solar Panel

Electric Current

Charge Controller

Inverter

(and/or)

Battery System

AC Power

DC Power

FIGURE 5.10 Conversion techniques of solar batteries (*How is Solar Energy Stored in 2019? | EnergySage*, 2020).

A grid-linked Photovoltaic system could help in reducing the capital and upkeep price by removing the requirement for battery storage. The grid could be turned to a storage system for a solar photovoltaic system, and this can also be withdrawn when necessary.

5.6.4 CENTRALIZED GRID-CONNECTED PHOTOVOLTAIC SYSTEMS

This system brings an alternative to the conventional power generation process and it strengthens the utility distribution system.

5.7 APPLICATIONS OF SOLAR ENERGY

5.7.1 SOLAR BATTERIES

Solar systems consist of panels, inverters, the equipment needed to install it on the roof, and the monitor for tracking of electricity generation. Panels gather the solar energy and convert it to electrical energy that passes via an inverter, which then converts it into another form, the kind that we can utilize to power our homes. These batteries store the energy generated via panels for future use. In addition, the battery has its own inverter and offers integrated conversion of energy. The amount of energy stored in the battery will be higher if it has more capacity. Figure 5.10 shows the conversion techniques of solar batteries.

5.7.2 SOLAR-PUMPING

Power production in solar pumping via solar energy is used for irrigation and household purposes. Its importance for aqua pumping is more evident in summer, paralleling the increase in sun-based rays; that is why this technique is more suitable for irrigation. Throughout the period of intemperate climate when the solar radiations are not sufficient to pump the water, pumping is also comparatively lesser then the transpiration loss from the harvests are also lower. Figure 5.11 shows a representation of solar-pumping in household work (Mondal, 2018).

5.7.3 SOLAR WATER HEATING

A sun-based water-heater includes a black FPC with the linked metallic tubes following the general route of the solar radiation, as shown in Figure 5.12. The collector has a cover of glass on top and a layer of insulation above it. The metallic tube of the collector is connected via a pipe to an insulating tanker that stores warm water in cloudy conditions. The collector does absorb the rays from the sun and transfers the heating factor to water flowing to the tube, either by gravitational force or via pumping. Hot water is provided to the storage tanker by the linked metallic tube. This system is mainly used in hotels, restaurants, etc. (Mondal, 2018).

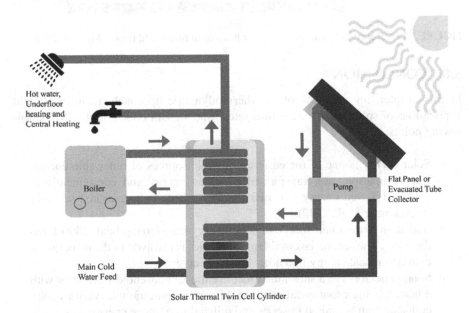

FIGURE 5.11   Representation of solar-pumping in household work using solar thermal twin cell cylinder (*Solar thermal panels | nidirect*, 2018).

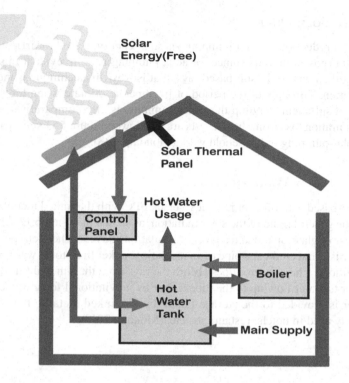

FIGURE 5.12 Representation of solar water heating in household tanks (Mondal, 2018).

5.8 CONCLUSION

In this chapter, an overview of the thermodynamic approach, techniques, and applications of solar energy were illustrated, and we can conclude it with the following points:

a. Solar energy is one of the easiest accessible sources of renewable energy. Lunar radiations are transformed into heat energy and consequently to electrical-driven energy via solar thermal-oriented concentrated schemes (Selvakumar et al., 2010).

b. Radiation is the main source of solar energy transferring heat (5800 K) to the earth. The second law of thermodynamics is involved in the principle of converting heat energy into other types of energy.

c. Solar collectors are a substitute for conventional household water heat with a heater, being economical over time. Even in domestic use, various solar collectors can be united in series and utilized to produce energy.

d. Solar panels gather the solar energy and turn it into electric energy that passes via an inverter; this is then converted into another form of energy. These batteries store the energy generated via panels for future usage.

REFERENCES

Al-Karaghouli, A., Renne, D., & Kazmerski, L. L. (2009). Technical and economic assessment of photovoltaic-driven desalination systems. *Elsevier*. https://www.sciencedirect.com/science/article/pii/S0960148109002626?casa_token=B0JnaAleJacAAAAA:5iJUqIhJjJTJGHc8bUDt5waRS0Snkn_bI_0DjusUQ3I5RgMnXFiJtB1N4i7nQHLow4khfSv7AeJV-w

Barbier, E. (2010). *A global green new deal: Rethinking the economic recovery*. https://books.google.com/books?hl=en&lr=&id=kDlLniTc97oC&oi=fnd&pg=PR7&dq=Babier+Edward+B.+A+global+green+new+deal.+1st+ed.+Nairobi, +Kenya:+UNEP%3B+2009.&ots=PeK9vAfwOq&sig=_2nRSiywK-266KmgSbox7pywbcM

Coskun, C., Oktay, Z., & Dincer, I. (2012). Thermodynamic analyses and case studies of geothermal based multi-generation systems. *Elsevier*. https://www.sciencedirect.com/science/article/pii/S0959652612001382?casa_token=aWo1yTrBn4oAAAAA:zNRfNF01id2Xhwc6sm7tReSlQxwK2L2jb2fouowp8MBpxTmL4yJYFTmhqHLV6y-V8xxfZ0uq6hMr_Q

De Vos, A., Landsberg, P. T., Baruch, P., & Parrott, J. E. (1993). Entropy fluxes, endoreversibility, and solar energy conversion. *Journal of Applied Physics, 74*(6), 3631–3637. doi:10.1063/1.354503

Gong, X., & Kulkarni, M. (2005). Design optimization of a large scale rooftop photovoltaic system. *Solar Energy, 78*(3), 362–374. doi:10.1016/j.solener.2004.08.008

How is Solar Energy Stored in 2019? | EnergySage. (2020). https://www.energysage.com/solar/solar-energy-storage/how-do-solar-batteries-work/

Kalogirou, S. A. (2013). *Solar Energy Engineering Processes and Systems*. https://books.google.com/books?hl=en&lr=&id=wYRqAAAAQBAJ&oi=fnd&pg=PP1&dq=Kalogirou+S.A.+Solar+energy+engineering+--+processes+and+systems.+2009.+&ots=LaI1D_JMMK&sig=tyOD4jU7y8MIJTtf4JcPd_osECc

Kennedy, C. (2002). *Review of mid-to high-temperature solar selective absorber materials*. https://www.osti.gov/biblio/15000706

Key World Energy Statistics 2020 – Analysis - IEA. (2020). https://www.iea.org/reports/key-world-energy-statistics-2020

Kikuchi, M., Lackner, K., & Quang, M. (2012). Fusion physics. *IAEA*, 24–26. https://courses.lumenlearning.com/physics/chapter/32-5-fusion/

Liu, S., Perng, Y., & Ho, Y.-F. (2014). The effect of renewable energy application on Taiwan buildings: What are the challenges and strategies for solar energy exploitation? *Elsevier*. https://www.sciencedirect.com/science/article/pii/S1364032113004632?casa_token=2ZUvBhJjjUYAAAAA:kxUJBzl3sUNSPav-y2WZiMbvG86zhQMA_Z8bj4FjdAUzN_083HXezlWkpnB5rdSNjIncNNB4nXCsLA

Liu, Z. (2015). *Global Energy Interconnection*. https://books.google.com/books?hl=en&lr=&id=T_MQCgAAQBAJ&oi=fnd&pg=PP1&dq=Liu+Z.+Global+energy+interconnection.+Supply+Demand+Glob+Energy+Electricity+2015%3B4:101–82.+&ots=zPNWlFY4l3&sig=EP60FAuJeR1p00KrdQj7wx8pTEY

Mills, D. (2000). Renewable energy in Australia. *Energy and Environment, 11*(4), 479–509. doi:10.1260/0958305001500257

Minardi, J., & Chuang, H. N. (1975). Performance of a "black" liquid flat-plate solar collector. *Elsevier*. https://www.sciencedirect.com/science/article/pii/0038092X75900572

Mondal, P. (2018). *Solar Energy: 10 Major Application of Solar Energy–Explained!* Your Article Library. https://www.yourarticlelibrary.com/energy/solar-energy-10-major-application-of-solar-energy-explained/28197

Morse, E., & Turgeon, A. (2012). *National Geographic Society - Solar energy*. https://www.nationalgeographic.org/encyclopedia/solar-energy/

Moshövel, J., Kairies, K. P., Magnor, D., Leuthold, M., Bost, M., Gährs, S., Szczechowicz, E., Cramer, M., & Sauer, D. U. (2015). Analysis of the maximal possible grid relief from PV-peak-power impacts by using storage systems for increased self-consumption. *Applied Energy*, *137*, 567–575. doi:10.1016/j.apenergy.2014.07.021

Nkwetta, D. N., Smyth, M., Van Thong, V., Driesen, J., & Belmans, R. (2010). Electricity supply, irregularities, and the prospect for solar energy and energy sustainability in Sub-Saharan Africa. *Journal of Renewable and Sustainable Energy*, *2*(2), 023102. doi:10.1063/1.3289733

Ozturk, I., Aslan, A., & Kalyoncu, H. (2010). Energy consumption and economic growth relationship: Evidence from panel data for low and middle income countries. *Energy Policy*, *38*(8), 4422–4428. doi:10.1016/j.enpol.2010.03.071

P Würfel. (2002). Thermodynamic limitations to solar energy conversion. *Elsevier*. https://www.sciencedirect.com/science/article/pii/S1386947702003557

Pekruhn, W., Thomas, L. K., Broser, I., Schröder, A., & Wenning, U. (1985). Cr/SiO on Cu solar selective absorbers. *Solar Energy Materials*, *12*(3), 199–209. doi:10.1016/0165-1633(85)90058-9

Phil, C. (2011). Solar energy perspectives. 30th ISES Biennial Solar World Congress 2011, SWC 2011, 1, 31–41.

Redaelli, G. (1976). Semitransparent tin-oxide films on Pyrex plates: Measurements of reflectivity. *Applied Optics*, *15*(5), 1122. doi:10.1364/ao.15.001122

Selvakumar, N., Barshilia, H., SE, K. R.-N. P. D., & 2010, U. (2010). Review of sputter deposited mid-to hightemperature solar selective coatings for flat plate/evacuated tube collectors and solar thermal power generation.

Seraphin, B. (1976). OPTICAL PROPERTIES OF SOLIDS: NEW DEVELOPMENTS. Opt Prop of Solids, New Dev. http://archives.umc.edu.dz/handle/123456789/118867

Solar thermal panels | nidirect. (2018). http://solarthermalpitsugoko.blogspot.com/2017/05/types-of-solar-thermal-panels.html

Weiss, W., & Rommel, M. (2008). IEA SHC-Task 33 and SolarPACES-Task IV: Solar Heat for Industrial Processes.

Würfel, P. (1988). Generation of entropy by the emission of light. *Journal of Physics and Chemistry of Solids*, *49*(6), 721–723. doi:10.1016/0022-3697(88)90206-5

Zahedi, A. (1997). *Energy: Concerns and possibilities*. https://research.monash.edu/en/publications/energy-concerns-and-possibilities

Zahedi, A. (2004). *The engineering and economics of solar photovoltaic energy systems*. https://researchonline.jcu.edu.au/11397/2/11397_Zahedi_2004_front_pages.pdf

Zaman, K., Tan, S., Sajjad, F., & Khan, M. A. (2014). Article in environmental science and pollution research. *Springer*, *21*(12), 7425–7435. doi:10.1007/s11356-014-2693-2

6 Memristive Behavior
Tool for Fault Detection and Repairing Dye Solar Cells

Manish Bilgaye, M. Gurunadha Babu,
and Y. David Solomon Raju
Holy Mary Institute of Technology
and Science, Roorkee, India

Adesh Kumar
University of Petroleum and Energy
Studies, Dehradun, India

CONTENTS

6.1 INTRODUCTION

Green energy is a synonym for sustainable development of ecology in which all forms of life thrive and evolve. PV basically converts the solar energy into electric energy in the most nonvolatile way as possible and, hence, is very popular as it pertains to a bright future. This chapter presents a brief review of PV technology and its development. The focus is on DSSC with insights on device physics, photogeneration, charge transportation, and elaboration of behavior of DSSC as a memristor – the fourth fundamental passive component. A brief introduction on memristors is included. Emphasis is on DSSC fault detection, isolation, and working of DSSC in faulty atmosphere using the memristor. A DSSC Spice model is developed, and a scheme for fault detection and repair, sensing operation, k-segment sensing, memristor resetting, and configuration and repair mechanism has been presented, followed by the conclusion and future scope of study.

6.1.1 GREEN ENERGY

Yamani Ahmed Zaki, oil minister of Saudi Arabia, has famously quoted "Stone age didn't end for lack of stone and oil age will end long before the world runs out of oil" [1] after understanding the harm being caused to the environment and by assessing the future of fossil fuels. The rapid growth of population is resulting in increase in energy demand. Also, there is limited capability to supply nonrenewable energy sources, mainly fossil fuels, and it has certain major disadvantages associated with it, such as environmental pollution and risk of climate change. However, there is sufficient coal to last over a century, and gas and oil reserves will last till the end of this century [2], but the demand estimate can be understood by calculating the yearly aggregate consumption of the sources of energy by the population of the world, yielding 113,009 TWh producing 25,606 TWh of electricity in 2017 and 117,837.2 TWh equivalent to 26,700 TWh in 2018 [3]. The contribution of various sources for the year 2018 is summarized in Table 6.1. Green and renewable energy is the energy of the future because it is from flow-limited naturally replenishing sources, thereby making it virtually inexhaustible over the time scale with a cap on the energy quantity obtainable over equal intervals of time.

TABLE 6.1

Contribution by Various Energy Sources

Sr. No.	Energy Source	Contribution (%)
1	Coal	38
2	Gas	23
3	Hydro and related	19
4	Nuclear	10
5	Solar and related	7
6	Oil and Gas	3

TABLE 6.2

Worldwide PV Data and CO_2 Reduction for Year 2019 and Estimate for Year 2020

Estimate for Year 2019

1	Worldwide PV installed capacity by the end of 2019 is 627 GW. This corresponds to 3% of world's electricity generation
2	Eighteen countries have installed 1 GW of PV Nine countries with minimum of 10 GW PV
3	Solar PV per capita Germany – 595, Australia – 585, Japan – 497
4	Countries with the highest PV penetration Honduras – 14.8%, Israel – 8.7%, Germany – 8.6% and India stands at the 8th position with 7.5% PV penetration.
5	Top PV market China – 30.1 GW, European Union – 16.0 GW and USA – 13.3 GW
	Solar electricity production avoided about 720 Mt of CO_2 emissions in the year 2019.
6	Capacity addition by the year 2020 end is estimated to 115 GW worldwide

The main renewable and green energy sources are solar, wind, geothermal, hydropower, rain, tides, waves, and biomass. Biomass can be further categorized to wood and discarded wood, waste in the form of municipal solids, biogas and landfill gas, biodiesel, and ethanol [4]. Contribution of solar photovoltaics is summarized in Table 6.2. There are four major areas where renewable energy is steadily replacing conventional sources of energy – heating of hot water and space, generation of electricity, fuels for vehicles, and energy requirements for rural and nonurban areas which are out of the reach of power grids. According to REN21's 2019 report, renewable energy meted a total of 18.1% of the world's energy consumption need and 26% of generation of electricity. Important issues like change in the world's climatic conditions, security in form of easy availability of clean energy, and fiscal benefits are being mitigated and addressed by the massive and swift deposition of highly efficient sources of renewable energy.

6.1.2 PHOTOVOLTAICS

Edmond Becquerel [5], also known as the father of photovoltaics, discovered photovoltaic effect in 1839. He observed that the light on an electrode dipped in acidic medium results in current through the electrode (Figure 6.1). He exposed the electrodes to blue light, ultraviolet, and sunlight.

In 1877, Adams and Day [6] reported the action of light on a bar of crystalline selenium, wherein its resistance was less when bar was exposed to light than it was when kept in dark. The selenium cell is represented through Figure 6.2.

In the year 1883, Fritts [7] proposed first thin film selenium solar photovoltaic cell, depicted in Figure 6.3. Current flows through the contacts to the external circuit. The PV cells were re-creatable and reproducible, but the underling theory was not clear.

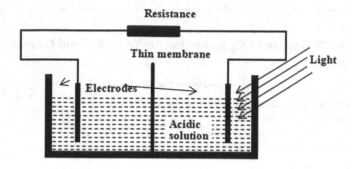

FIGURE 6.1   Becquerel's experimental setup.

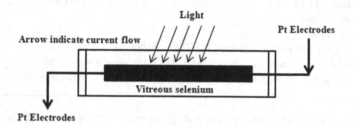

FIGURE 6.2   Selenium photovoltaic cell.

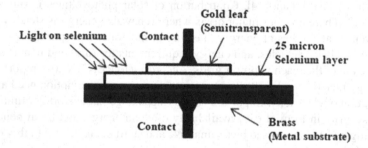

FIGURE 6.3   First thin-film solar photovoltaic cell.

The year 1900 saw the birth of quantum mechanism proposed by Max Plank [8], wherein he introduced concept of energy through the equation $E=hv$. Here, E represents Quanta, a discrete quantized energy packet proportional to frequency represented by v and h = Plank's constant. He reached this conclusion when he was not able to solve the problem related to thermal radiation by means of traditional classical physics. In 1905, Albert Einstein published wok on photon packets, i.e. light quanta called photon. Through this, he completely laid down the principles of photovoltaic mechanism and the foundations of the semiconductor industry. In 1933, Grondahl [9] described the details behind the manufacturing process of the photoelectric solar cell. It became popular owing to the low manufacturing cost involved. In 1941, Ohl filed the first patent [10] for silicon solar cell. It had an efficiency of less than 1%. Despite it not carrying commercial value, it was a landmark point in history.

In 1954, Chapin et al. [11] proposed a silicon semiconductor solar cell having an efficiency of 6%. This was an offshoot of the observation in which silicon diode produced a significant amount of current and voltage in the presence of light. It was commercialized and can be thought as the beginning of silicon solar cells. After this, the technology saw rapid improvements with a host of semiconductors developed, as summarized in Table 6.3.

Early solar cells were used in telephone repeaters. In 1958, six solar cell panels were first used in space application over the satellite Vanguard 1. It produced a power of 5 mW and was functional for 6 years after the battery was discharged. Nowadays, solar panels have been well understood and accepted, resulting in applications like roof-top solar panels. On July 10, 2020, the Prime Minister of India Mr. Narendra Modi inaugurated the Rewa solar power plant, the largest in Asia having a production capacity of 750 MW. Solar power is poised to become a major source of electric supply for this planet.

6.1.3 PV CELL BEHAVIOR AND V–I CHARACTERISTICS

PV cell characterization is important from the viewpoint of the analysis and design of PV-based systems. The PV cell is similar in behavior to a *p-n* junction diode. The incident light raises the valence band gap to the conduction band. Current direction

TABLE 6.3
Summary of Photovoltaic Solar Cell Technology

Group	Photovoltaic Solar Cell Technology	Efficiency (%)
1	(Mono Si) Monocrystalline Silicon	20
	(Multi-Si) Polycrystalline	18
	(TFSC) Thin Film	18

Note: Group 1 are the mainstay in solar cell as of now.

2	Gallium Arsenide Germanium	30
	Copper Indium Gallium Selenide	21
	Cadmium Telluride (CdTe)	21
	(a-Si) Amorphous Silicon	10
	(DSSC) Dye Synthesized	11

Note: Promising next-generation solar cell technology.
Amorphous silicon promises very low production cost.

3	(OPV) Organic solar cell	8
	Multi-junction solar cell (InGAP/GaAs/InGaAs)	37

Note: Organic solar cell has low production cost.
Multi junction solar cell claims very high efficiency.

4	Perovskite	Upcoming
	Quantum dot	Solar Cell
		Technology

Note: Perovskites materials were initially used as a coating material for silicon solar cell and were instrumental in a hike of efficiency by 5%.
Quantum dots claims high efficiency as solar cell.

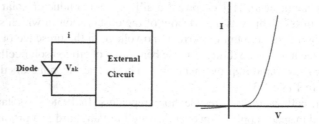

FIGURE 6.4   PV cell behavior as a sink.

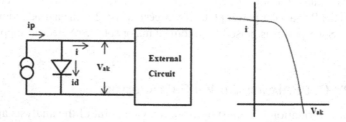

FIGURE 6.5   PV cell behavior as a source.

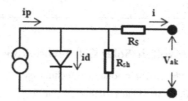

FIGURE 6.6   PV cell equivalent model.

indicates dissipation of power in the diode making it behave like a sink, as shown in Figure 6.4.

However, the desired behavior of the PV cell must be like that of a source, and this is possible in the scenario depicted in Figure 6.5 wherein the current i_d is flowing from anode to cathode and excess current 'i' is flowing to the external circuit. This is because of the photon current 'i_p', produced by the PV cell upon exposure to sunlight. The terminal potential is 'V_{ak}', and its direction is also retained. Under dark conditions, the characteristics are equivalent to those of a simple diode but on exposure to solar intensity the V–I characteristics are as described in Figure 6.5. V–I characteristics of solar cell indicate a unique feature of being both the current and voltage source at different instances of time. R_{sh} and R_s represent the equivalent shunt and series resistances of these sources, respectively. The PV cell model is represented in Figure 6.6 where i_p is the photocurrent, and shunt and series non-idealities have been shown by R_{sh} and R_s, respectively.

The equation for terminal current and voltage is

$$i_p = i_d + iR_{sh} + i \qquad (6.1)$$

$$VR_{sh} = V + iR_s \tag{6.2}$$

The diode current is represented by

$$i_d = I_0 \left(e^{\frac{v+iR_s}{R_{sh}}} - 1 \right) \tag{6.3}$$

Here,

$$V_T = \frac{kT}{q} = \frac{T}{11,600} \tag{6.4}$$

$I_0 = KT^m\, e^{-VG_o/\eta v_T}$ = Reverse saturation current dependent on material, doping, and temperature

K = Constant dependent on dimension and material property

VG_o = Numerical equivalent bandgap energy of electron in ev

m = 1.5 for silicon

VG_o = 1.16 to 1.21 for silicon. Value is dependent on grade and purification.

Therefore,

$$i = i_p - I_0 \left(e^{\frac{v+iR_s}{\eta v_T}} - 1 \right) - \frac{v + iR_s}{R_{sh}} \tag{6.5}$$

Equation (6.5) represents the terminal current model of PV cell represented in Figure 6.6.

The variation in V–I characteristics is explained through Point 1, which represents the short-circuit point in PV cell V–I characteristics wherein $V_{ak} = 0$, $R_s \ll R_{sh}$ so Equation (6.4) becomes $I_{sc} = i_p$. This condition represents incident solar power and is known as insolation. Point 2 represents the open-circuit point wherein $v = V_{oc}$ and $i = 0$. For these conditions, Equation (6.4) becomes $V_{oc} = nV_T \ln\left(\frac{i_p + i_o}{i_0} \right)$. Here, V_{oc} is related to insolation logarithmically, indicating the change in incident solar power changes the I_{sc} linearly, as represented in Figure 6.7b. $p = vi$ represents the power curve and point 3 represents peak power.

Solar cell efficiency:

$\eta = \dfrac{P_o}{P_{in}}$ is cell efficiency at peak power P_m, wherein $P_m = \dfrac{V_m I_m}{P_{in}}$.

$$P_{in} = 1 \left(\frac{kW}{m^2} \right) \times (\text{Solar Panel Area})$$

6.1.2.2   Influence of Temperature on Solar Cell Characteristics

Effect of temperature on short-circuit current I_{sc}: With increase in temperature, there is reduction in the bandgap energy, allowing more electrons from the valence band to cross to the conduction band, resulting in extra photon current, thereby increasing I_{sc}. Typical change is 0.1%/°K for silicon.

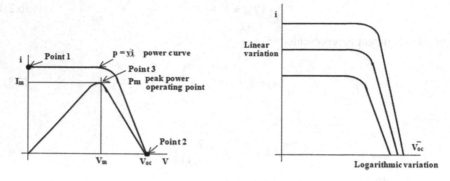

FIGURE 6.7 (a) Solar PV cell $V–I$ characteristics representing open-circuit, short-circuit, and maximum power points. (b) Linear versus logarithmic relationship between I_{sc} and V_{oc}.

Effect of temperature on open-circuit voltage V_{oc}:

$$V_{oc} = \eta V_T \ln\left(\frac{i_p + I_o}{I_o}\right) \tag{6.6}$$

Here: $\eta = 2$ for silicon, V = volt equivalent of temperature.
For the condition $I_o \ll I_p$,

$$V_{oc} = \eta V_T \ln\left(\frac{i_p}{I_o}\right) \tag{6.7}$$

Also,

$$I_o = KT^m e^{-V_{Go}/\eta V_T} \tag{6.8}$$

$$\ln\{I_o\} = \ln\left(KT^m\right) + \ln\left(e^{\frac{-V_{Go}}{\eta V_T}}\right) \tag{6.9}$$

$$\ln\{I_o\} = m\ln(KT) + \left(-\frac{V_{Go}}{\eta V_T}\right) \tag{6.10}$$

$$\ln\{I_o\} = m\ln(K) + m\ln(T) - \frac{V_{Go}}{\eta T / 11,600} \tag{6.11}$$

Differentiating with respect to T

$$\frac{d}{dT}(\ln I_o) = \frac{m}{T} + \frac{V_{Go}}{\eta T V_T} \tag{6.12}$$

Here $V_{co} \propto \dfrac{1}{T}$ and V_{Go} is numerical equivalent of bandgap energy.

From Eq. (6.7),

$$V_{oc} = \eta V_T \ln\left(\frac{i_p}{I_o}\right) \tag{6.13}$$

$$\frac{V_{oc}}{\eta V_T} = \ln(i_p) - \ln(I_o) \tag{6.14}$$

Differentiating with respect to T and considering variation of $i_p \ll I_o$

$$\frac{d}{dT}\left(\frac{V_{oc}}{\eta V_T}\right) = \ln(i_p) - \ln(I_o) \tag{6.15}$$

$$-\frac{V_{oc}}{\eta V_T} + \frac{1}{\eta V_T}\frac{dV_{oc}}{dT} = -\left(\frac{m}{T} + \frac{V_{Go}}{\eta T V_T}\right) \tag{6.16}$$

Hence, we get

$$\frac{dV_{oc}}{dT} = \frac{V_{oc} - (V_{Go} + m\eta V_T)}{T} \tag{6.17}$$

For silicon at 300°K, $m = 1.5$, $\eta = 2$, $V_{Go} = 1.16$, $V_{co} = 0.6$,

$$\frac{dV_{oc}}{dT} = -2.12 \; mV/_{°K} \tag{6.18}$$

indicating $V \propto \dfrac{1}{T}$.

Effect of temperature on power:

$$Power = v \times i \tag{6.19}$$

The coefficient of current and voltage is positive and negative, respectively, effectively giving a negative coefficient of power.

6.2   DYE-SYNTHESIZED SOLAR CELL

Physical processes, mainly generation of photons, disassociation of excitons, and transportation of charge, are at the center of working of DSSC. The framework for amalgamation of these properties involves customized materials and design of their junctions. A conductive oxide glass substrate is deposited with approximate thickness of few tens of microns n-type but transparent and wide band semiconductor having configuration of either nanoscale network of nanoparticles [12], nanotubes [13] or nanofibers [14] which ultimately forms a photoanode. The primary objective of the photoanode nanostructure is to provide a significantly big surface area to house a single layer of dye activator/sensitizer. The dye sensitizer does the function of absorption of the light. A channel for transfer of charge between the dye and counterelectrode is disseminated through a solution of an electrolyte consisting of a redox couple.

Under illumination, an exciton is produced by the excited dye, which separates at the junction of the semiconductor and the dye and enters the semiconductor's conduction band and further traverses through the DSSC photoanode nanostructure network to the external load. The molecules of dye (oxidized) bear holes (e^+) and are rejuvenated by the electrolyte's reducing class as shown in Figure 6.8. It confirms the continuous creation of reducing species $\left(I_3^-\right)$ at the counterelectrode and furnishing of the electrons to the dye molecules. For effectively harvesting the light, the dye mandates possession of a broad absorption range – near-infrared to visible stretch of the spectrum of the solar rays. It can be observed that the DSSC system encompasses various interfaces (unconventional) like solid–solid and solid–liquid, owing to its structure involving many bulk materials. Also, the kinetics related to the charge transfer varies considerably, making it difficult to build an all-inclusive and broad-ranging mathematical model that may comprise all intricacies, thereby eluding a holistic explanation. It necessitates the study of numerous physical procedures distinctly and in great detail.

6.2.1 Overview of Device Physics

The photons are absorbed by the sensitizer and converted to excitons. At the dye–semiconductor interface, the diffused excitons split to holes and electrons owing to affinity dissimilarities of the electron. This parting entails the dye's lowest unoccupied molecular orbital (LUMO) to be appropriately higher than the semiconductor's conduction band so that the exciton splits. The diagrammatic representation of the energy is depicted in Figure 6.9, which illustrates that the photovoltaic act of DSSCs rises from an inherent operational field reached by the affinity dissimilarities of the

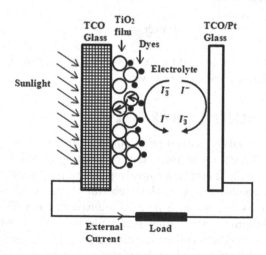

FIGURE 6.8   Structure of the dye-synthesized solar cell [12].

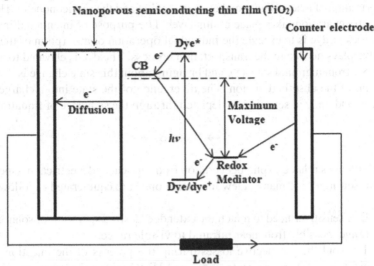

FIGURE 6.9 Energy diagram of dye-synthesized solar cell depicting direction of electron flow.

electron than generally thought of as an in-built electrostatic field reached by the work function difference. Band bending is considered as the source of photovoltage in the DSSC [15] and is prevalent in both the mechanisms.

The disassociation of the exciton [16] at the semiconductor–dye interface takes place by the injection of the electrons into the semiconductor, causing holes to remain back in the dye that are continually evacuated by the electrolyte's reducing species. The process of separation of charge is very effective and occurs in an ephemeral time period of picoseconds (10^{-9}). The foremost loss mechanism occurs to the injected electrons when they cross/route through the grid of the semiconductor to the contact in the front. Because of the sluggish passage in the film of the semiconductor, the photo-generated electrons are expected to recombine with the holes localized in the dye and the electrolyte's oxidizing species.

The concentration of the charge carrier is controlled by the rate of the recombination, which is also instrumental in determining the density of the current. Generally, for DSSC in open-circuit condition, the rate of recombination matches the rate of photogeneration.

6.2.2 Photogeneration

The production of the exciton is due to an excited state, which is the resultant of the process termed as photogeneration that involves absorption of a photon to reach to the excited sate. The exciton further disintegrates into a pair: an electron and a hole. However, the generation of mobile charges is not an automated process and does not depend solely on the excitonic absorption occurrence; hence, it mandates inclusion of the phenomenon of exciton generation and disassociation to calculate the agreeable free electron generation rate.

A single layer of dye is fashioned on the surface of the semiconductor, resulting in the formation of a heterojunction between the dye and the semiconductor. Hole–electron pair separation takes place at this layer. The purpose of intentional insertion of molecules of dye is to execute the individual operation of absorption of the photons. The dye plays no role in the transportation process. The dye is elevated to an excited state (S^*) from a ground state (S), and bringing about this state change is the primary function of the absorbed photon. The divergence of the state into a charged pair of electron and ionized sensitizer is depicted through the sequence of Equation (6.20):

$$S \rightarrow S^* \leftrightarrow S^+ + e^- \tag{6.20}$$

The challenges related to photogeneration from optical and electrical perspectives on the dye sensitizer are many; a few important ones are enumerated as follows:

- Dye sensitizer need to match the extended desired spectrum of solar energy range, possibly from near infrared to visible range.
- To avoid the prospective losses during the process of the injection of the electrons, it is necessary to align the LUMO within the periphery of semiconductor's conduction band.
- The uppermost complete (full) orbit of the molecule needs to be of lesser potential compared with the potential of the electrolyte's redox in order to obtain the donation of the electron from an electrolyte in liquid form or from a hole conductor material.
- From stability and chemical structure organization viewpoint, the dye needs to carry supplementary requirements like possessing groups of anchors to fasten on the semiconductor.
- Must display stability over the duration of its life and meet the life expectancy of over two decades.

It is observed that the rate of photogeneration is determined primarily by the capacity of absorption of the light by the dye sensitizer.

6.2.3 CHARGE TRANSPORTATION

Apart from photogeneration, charge transport mechanism through liquid electrolyte-saturated semiconductor films needs thorough deliberation. Few important factors that need consideration are as enumerated:

- Candidate class for transportation of charge, namely ions, electrons, and cations.
- Major mechanism for transportation of charge, namely diffusion and drift in solids and liquids. Parameters that require investigation for their role are
  - Electric field
  - Electrolyte's screening effect
  - Charge carrier's spatial distribution

- Probable losses that may occur at respective hetero-junctions and interfaces, also known as contacts
- Band alliance of respective layer in diverse circumstances, mainly short-circuit, open-circuit, and maximum working points of power.

A universal approach instrumental to shed light on the transport mechanism prevailing in semiconductors is amalgamated to a set of three equations comprising the following:

- Diffusion and drift equations: The source of drift is the applied electric field, while the source of diffusion is the concentration gradient owing to the electron. Boltzmann's transport equation is helpful in the derivation of both the equations.
- Continuity equation: It defines the norms related to conservation of charge.
- Poisson's equation: This equation links, at any specific point in the semiconductor, the electric potential to the concentration of the charge.

6.2.4 MEMRISTOR

System performance improvement is a natural never-ending desire instrumental in driving the technological efforts for the development of next-generation high-performance systems. Currently, memristor-based ReRAM (Resistive Random Access Memory) has proven itself powerful through its potential as a leading candidate to mitigate issues on multiple technological fronts. Devices based on memristive ReRAM when coupled with a complementary metal oxide semiconductor (CMOS) transistors have numerous openings related to furtherance of next-generation high-performance systems [17,18]. Basically, memristor is a variable resistance device based on the relationship between charge 'q' and the flux 'ϕ' with the capacity to retain the previous resistive value based on amount of charge 'q' that has flown through it, in the process providing it the most important memory effect, that of nonvolatility in nature. The resistance value/resistive memory can be programmed to attain a particular value, which can be understood as logical binary bits '0' and '1', respectively. The second characteristic behavior of memristor is akin to the synapse of a brain [19].

Leon Chua postulated the concept of memristor [20], shown in Figure 6.10, which was later realized as a product by HP Labs [21] in 2008, as shown in Figure 6.11. The HP Labs memristor comprises a solitary layer of insulating (TiO_2) and a conductive metallic (TiO_2-x) layer sandwiched amid the electrodes made of platinum (Pt). These devices can be made up through better quality layouts via non-lithographic procedures similar to imprint lithography producing ultrahigh density memory, characterizing nonvolatile state retention (memory effect) capacity and having very small access latencies, enumerating the two parameters specifically vital for steady and stable storage of data. This emphasizes the high importance of designing and implementing the memories using memristor.

Two terminal memristor can attain in minimum two states 0 and 1 representative of high and low resistance respectively. These states can be changed upon application

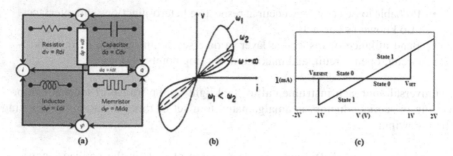

FIGURE 6.10 (a) The concept of memristor. (b) Generic memristor *V–I* characteristics. (c) *V–I* characteristics defining the states.

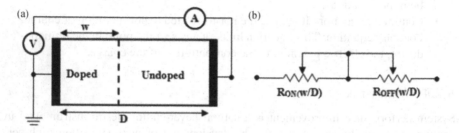

FIGURE 6.11 (a) HP Lab Memristor representation. (b) Circuit equivalent of HP Memristor.

of appropriate voltage values, which can be understood from its *V–I* characteristics depicted in Figure 6.10. R_{ON} and R_{OFF} represent the memristor's ON and OFF resistances values, and $i(t)$ and $v(t)$ represent the current through and the voltage across the memristor, respectively, as per Figure 6.10c. General equations representing symmetrical highly nonlinear *V–I* characteristics and programming thresholds are as follows:

$$\frac{dw}{dt} = \mu_v \frac{R_{ON}}{D} i(t) \tag{6.21}$$

$$v(t) = \left(\frac{w}{D} R_{ON} + \left(1 - \frac{w}{D}\right) R_{off}\right) i(t) \tag{6.22}$$

Some prominent applications of memristor are as follows:

- Efficient and effective nonvolatile storage for
  - Systems based on binary number logic
  - Systems based on multi-bit system
- Major design constituent for
  - Analog and digital circuits [22]
  - Logic synthesis [23]
  - Implications-based logic [24]
  - Hybrid circuits [25]

- An additional promising use of memristor is the design of content address-able memory (CAM) [26]. This happens to be a different sort of memory finding use in search applications in computer networking structure for sev-eral search roles such as
  - Lengthiest prefix matching
  - Classification of multi-field packets
  - Caches of the processors
  - Neuromorphic hardware

Many of the applications cited are indicative of the emergence of memristor's potential on the new technology arena. It poses as an alternate promising candidate technology having the potential to turn over the restrictions of conventional CMOS centered arrangements of the system.

6.3 DSSC FROM MEMRISTIVE VIEWPOINT

DSSC fits in the green energy system roadmap since the technology is an able com-petent and prominent alternative to solar cell technology based on silicon [27–28], importantly at low cost [29–30]. Also, it exhibit high energy conversion efficiency; with the development and acceptance of the technology, it has penetrated to everyday use compared to its restricted use for critical applications earlier. The application range of DSSC is determined by the continuous operation of the PV cell in the pres-ence of faulty conditions. Reliable designs using DSSC are possible since it exhibits memristive behavior, thereby opening the opportunity to explore continuous moni-toring of the health of the PV cells in highly distributed mechanisms. Also, DSSC use roughly similar material and techniques for design as that used by the memristor. The primary advantages are creation of scalable on-chip solution for detection of fault(s), provision of scope to attain maximum advantage to integrate hybrid CMOS/ memristor technology to open a window for planning detection of the faults, and building reconfiguration blocks as autonomous entities.

6.3.1 DSSC FAULT DETECTION

Isolating the fault after finding it is the crucial central key aspect, independent of the type of PV system put into use. Flawed production technique and alignment contrib-ute to defects. Elementary origins may be cited to excessive exposure to the heating process, unclean production environment cause to contaminations, doping imper-fections, diverse rates of crystallization, active region being compromised with the presence of the unwanted impurities to name a few. The family of faults of solar PV cell constitutes of permanently fixed unhealthy or semi-healthy cells generating fully or partially degraded cell voltages and currents. One of the most common but critical faults in PV systems is the ground fault, which owes its origin to chance near to the zero resistance route for current, to the real ground terminal of the system. Short- and open-circuit PV cells are other prominent faults. These faults are responsible for deviation of the voltage from its normal value and are reflected at the node of the array of the PV cells. The technique utilizes voltage measurement as a tool to figure out the fault to form the basis for PV cell array reconfiguration but based on the supposition of existence of reconfiguration switches and redundant parallel building blocks as such in the global system.

Figure 6.12 compares the difference in cell behavior through V–I characteristics of a DSSC PV system comprising of four cells in faulty and healthy conditions, respectively [31]. The ground fault is incorporated in one out of four PV cells. Figure 6.13 points to the shift in the maximum power point. This indicates the necessity to prioritize smart detection of faults for efficient recovery for systems based on DSSC.

6.3.2 DSSC SPICE MODEL

Basic bottlenecks are cost and challenges on technical front, making them primary concerns for different PV cell technologies, like PV based on wafer silicon, thin film-based PV, and concentrated photovoltaic (CPV) as being competitive with customary sources of power.

$$j = j_0 \left(e^{\beta \left(e_0 / KT \right) V} - e^{-(1-\beta) \left(e_0 / KT \right) V} \right) \tag{6.23}$$

However, the technically evolved third-generation PV cell technology, particularly the DSCC is proving to be a viable option from multiple perspectives. Circuit equivalent modeling of a single cell of DSSC using GVALVE block is possible,

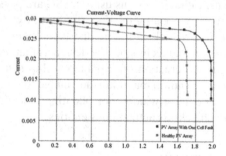

FIGURE 6.12   Single faulty cell's effect on V–I curve.

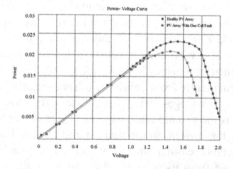

FIGURE 6.13   Single faulty cell's effect on the maximum power point tracking.

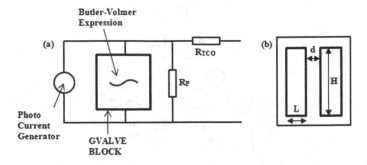

FIGURE 6.14 (a) DSSC PSPICE circuit. (b) DSSC typical layout module.

which allows the scope for describing $J-V$ characteristics through the diode current equation [32–33]. Equation (6.23) is the Butler–Volmer expression, and it provides ease and is instrumental to model the DSSC $V-I$ characteristics [34]. The simulation configuration constitutes of parallel configuration of a resistance (RP), current generator, and the GVALVE block, which is further connected in series with RTCO (resistance of transparent conducting oxide) as depicted in Figure 6.14. Current generator is responsible for the simulation of the photocurrent [29].

6.3.3 DSSC FAULT DETECTION AND REPAIR USING MEMRISTOR

Figure 6.15 shows a universal representation of the design for the detection of the fault using memristor/CMOS hybrid platform as a central theme. The idea is based on the usage of surplus redundant cells and a module for the detection of the faults. Probable behavior, indicating a fault within the system, is attributed to the manner in which the change of the state takes place and is captured in form of time delays. However, apart from memristor/CMOS hybrid platform, the block also consists of a standard configuration that includes an array of solar cells and load supplied by DC–DC converter. Figure 6.16 is the circuit that senses the voltage and further relates it to the width of the pulse and behavior of the change of the state of the memristor as follows:

Supposing voltage at input and C is constant, then the memristor switching time is

$$T = \frac{C}{R_{\text{off}}} \tag{6.24}$$

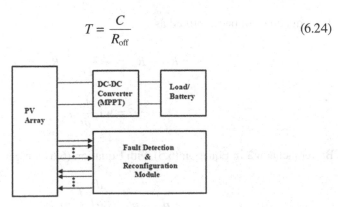

FIGURE 6.15 PV system fault detection module.

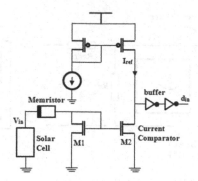

FIGURE 6.16   Circuit for single cell fault detection.

The assumption is under the supposition that initially transistor $M2$ is either switched off or working near the linear region, i.e., $(Vgs < Vth)$. The change of the state of $M2$ begins at time reference $T1$, where it starts to conduct the current equivalent in magnitude to I_{ref} and culminates at time reference $T2$, wherein $M2$ enters the saturated state/region.

$$\text{At } T1 \begin{cases} \text{Memristor state} = w1 \\ \text{Memristance} = Rw1 \end{cases}$$

and

$$\text{At } T2 \begin{cases} \text{Memristor state} = w2 \\ \text{Memristance} = Rw2 \end{cases}$$

resulting in the following:

$$R_{w1} - R_{w2} = \frac{w1}{D} R_{on} + \left(1 - \frac{w1}{D}\right) R_{off} - \left(\frac{w2}{D} R_{on} + \left(1 - \frac{w2}{D}\right) R_{off}\right) \quad (6.25)$$

The equation can be simplified as

$$R_{w1} - R_{w2} = \frac{w2 - w1}{D} R_{off} \quad (6.26)$$

$$w2 = w1 \frac{dw}{dt} T \quad (6.27)$$

By replacing $w2$ in Equation (6.5) with Equation (6.6), we get

$$R_{w1} - R_{w2} = \frac{\dfrac{dw}{dT} T}{D} R_{off} \quad (6.28)$$

For R_{w1}, (time $T1$), voltage V_{GS2} for $M2$ is marginally higher than V_{th} (we estimate $V_{GS2} - V_{th} \approx 0.1$ V), thereby closing $M2$ and allowing current flow through it, leading to the new equation for R_{w1} as follows:

$$R_{w1} = \frac{V_{in} - V_{th}}{(0.1)^2 \, \dfrac{\mu \, C_{ox}}{2} \, \dfrac{W}{L}} \qquad (6.29)$$

For R_{w2}, (time $T2$), $M2$ operates in a saturated state, making reference current I_{ref} to flow via $M2$. $M1$ copies the reference current to the memristor, leading to the equation for R_{w2} as follows:

$$R_{w2} = \frac{V_{in} - \left(\sqrt{\dfrac{I_{ref}}{\dfrac{\mu \, C_{ox}}{2} \dfrac{W}{L}}} + V_{th} \right)}{I_{ref}} \qquad (6.30)$$

dw/dT as in Equation (6.7) is a function representing the voltage drop across the memristor. Also, dw/dT is not a constant quantity since potential drop is linear at time instances $T1$ and $T2$, respectively. Hence, Equation (6.7) takes the form

$$T = \frac{C}{R_{off}} \qquad (6.31)$$

where C is calculated from Equations (6.5) to (6.9) and assumes a constant value.

Associated assumption, wherein R_{off} is a constant, the switching time of the memristor and hence the width of the output pulse is proportional to input voltage V_{in} and behavior is nonlinear in nature. Substituting Equation (6.8) and (6.9) representing $Rw1$ and $Rw2$, respectively, in Equation (6.7), the relationship of V_{in} with T takes the form

$$A - BV_{in} = \frac{dw}{dt} T \qquad (6.32)$$

Here,

A and B are functions of I_{ref} and R_{off} and the parameters of the transistor
    $dw/dt = $ fn (drop of voltage across the memristor)
    As V_{in} increases, the memristor voltage increases, reflecting as a nonlinear rise of dw/dt, affirming the nonlinear relationship between T and V_{in} [35] as per Equation (6.11).
    The expression represented by Eq. (6.33) approximates the programmed threshold nonlinear V–I characteristics of a particular memristor type:

$$I = c \times \sinh(\alpha V(t)) \qquad (6.33)$$

Here, c and α are constants dependent on the state and instrumental in characterizing the memristor's state.

6.3.4 SENSING OPERATION

Detection of fault revolves around sensing the voltage and is realized through several sections consisting of current comparator and memristor. Each segment consists of single current source and the bulk and drain material is shared by the segments' mirrored transistors. Figure 6.16 shows the mechanism to sense the voltage for an individual cell, which is further used to evaluate the evolving characteristics of the state. R_{off} and parasitic capacitance of the transistor are responsible for the output delay [36]. One possible outcome, out of the following three, will happen during the operation pertaining to sensing of the voltage:

- For zero input voltage, the output (d_{in}) remains at logic '1'
- The output may switch to logic '0' in spite of the input voltage being normal
- Fault degrades the input voltage, resulting in output undertaking the transition to logic '0' but delayed in the time domain.

Ex-Or operation is performed on the signal output received from reference current and enables a signal that generally attains logic 1 state at the beginning of the measurement cycle, thereby generating information in the form of a pulse d_{in} representing the state transition phenomenon. This pulse is instrumental in determining the cell's health and needs calibration to identify healthy and faulty cells.

6.3.5 K-SEGMENT SENSING AND MEMRISTOR RESETTING

A programmable current reference along with several memristors is used to test the nodes. Also, this circuit arrangement helps in reducing the quantity of cycles required for the measurement purpose, as is shown in Figure 6.17a. Deviation between the expected or the standard delay and the actual delay throws light on the pattern of delay at the node and is instrumental in providing information about the fault. Careful design of the delay elements is necessary to match the pattern of delay. This generates a unique output code upon turning on the predefined quantity of segments. Thus, timing data proves vital for detection of different faults with a caution for meticulous calibration of information pertaining to the delay to ensure avoidance of untrue positives.

Memristor necessitates resetting once the measurement cycle is over or prior to the beginning of a second measurement cycle. Resetting the memristor to the initial state R_{off} is carried out with the help of the charge accumulated in the capacitor C_p. The reset circuit is shown in Figure 6.17b in which upon switching reset to logic 1 the potential across the memristor attains the value $-V_p$, which is also the capacitor voltage across the capacitor. Figure 6.18 is the representation of the memristor reset circuit. When reset pin is made 1, the capacitor voltage V_p reflects across the memristor in a reverse fashion, i.e., $-V_p$ in the course of transition of the state. Appropriate switch arrangement isolates the memristor and the PV system from each other during the reset phase. Individual testing of the segments of the solar cells results in linear increase in the time required for the testing purpose. In large and complex systems, the test stage may consume relatively very high time.

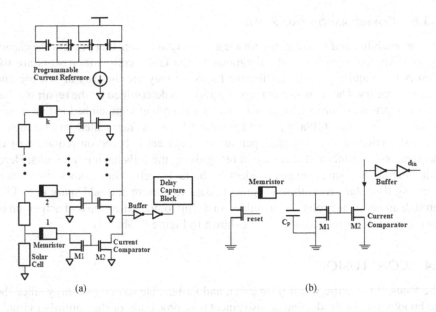

FIGURE 6.17 (a) Representation of k-segment comprising programmable current mirror circuit. (b) Memristor reset circuit.

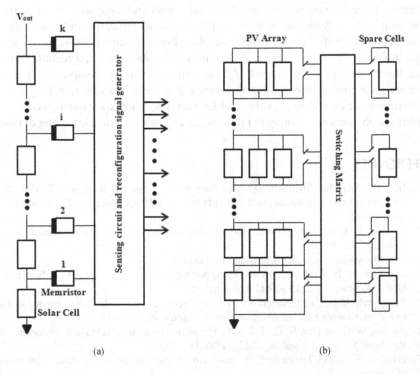

FIGURE 6.18 (a) Memristor sensing mechanism in solar cell array. (b) PV system and reconfiguration mechanism.

6.3.6 CONFIGURATION AND REPAIR

Sensor reliability and segment magnitude are the vital factors that possess the capacity to affect the accuracy of the diagnosis of the faulty cells. Memristors are the sensors that require periodic testing for faults for they are the part of sensing and control circuitry. The sequence of reconfiguration is determined by the repair mechanism comprising of memristor as sensor and a decoding logic embedded on a chip as depicted in Figure 6.18a, wherein the solar PV cell is fragmented into segments, and distribution of solar PV cells per sensing segment is based on requirement of the area, testing time, and accuracy in recognizing the cells having the faults. Here, switching mechanism is the key to identify the faulty cells. Numerous switches surrounding the solar cell within a segment adapt to remove and add spare cells [37] through an assistance solution embedded in a chip for identification and activation of the mechanism for reconfiguration, as shown in Figure 6.18b.

6.4   CONCLUSION

The focus of the entire world is on green and sustainable source of energy since the methodology of its production is instrumental in protection of the natural environ − the most prominent need of the hour. Photovoltaics is steadily moving to prominence in the green energy zone. Memristor is a parallel emerging technology that is steadily disrupting the conventional foundations of electronics and computing engineering in innumerable areas. A distinctive feature of this is the identification of the memristive phenomenon in DSSC PV cell, which is explored in this chapter, for identification of faults in DSSC PV cells, isolation of the faulty cells, its repair, and reconfiguration, while the PV module is still at work. A DSSC spice module is exemplified. However, there are issues owing to connection of solar PV modules with mismatching electrical characteristics spiraling to problems like irreversible breakdown in reverse-biased conditions, phenomena resulting in hotspots, and disproportionate lessening of power.

REFERENCES

1. IEA PV Snapshot 2019.pdf. *International Energy Agency.* Retrieved 2 May 2020. https://iea-pvps.org/wp-content/uploads/2020/01/Press_Release_T1_15042019_-_Snapshot.pdf
2. Shafiee, S., & Topal, E. (2009). When will fossil fuel reserves be diminished? *Energy Policy, 37*(1), 181–189.
3. https://www.iea.org/topics/world-energy-outlook
4. Murdock, H. E., Gibb, D., André, T., Appavou, F., Brown, A., Epp, B., & Sawin, J. L. (2019). Renewables 2019 global status report.
5. Becquerel, A. E. (1839, November 21). On electric effects under the influence of solar radiation. *Comtes Rendus de l'Academie des Sciences, 9,* 31–33.
6. Adams, W. G., & Day, R. E. (1877). V. The action of light on selenium. *Proceedings of the Royal Society of London, 25*(171–178), 113–117.
7. Fritts, C. E., (1883, December). A new form of selenium photocell. *American Journal of Science, 26,* 465–472.
8. Planck, M. (1900). On the theory of the energy distribution law of the normal spectrum. *Verh. Deut. Phys. Ges, 2,* 237–245.

9. Grondahl, L. O. (1933). The copper-cuprous-oxide rectifier and photoelectric cell. *Reviews of Modern Physics*, *5*(2), 141.

10. Ohl, R. S. (1948). U.S. Patent No. 2,443,542. Washington, DC: U.S. Patent and Trademark Office.

11. Chapin, D. M., Fuller, C. S., & Pearson, G. L. (1954). A new silicon p-n junction photocell for converting solar radiation into electrical power. *Journal of Applied Physics*, *25*(5), 676–677.

12. Gong, J., Liang, J., & Sumathy, K. (2012). Review on dye-sensitized solar cells (DSSCs): Fundamental concepts and novel materials. *Renewable and Sustainable Energy Reviews*, *16*(8), 5848–5860.

13. Sigdel, S., Elbohy, H., Gong, J., Adhikari, N., Sumathy, K., Qiao, H., ... Qiao, Q. (2015). Dye-sensitized solar cells based on porous hollow tin oxide nanofibers. *IEEE Transactions on Electron Devices*, *62*(6), 2027–2032.

14. Gong, J., Qiao, H., Sigdel, S., Elbohy, H., Adhikari, N., Zhou, Z., ... Qiao, Q. (2015). Characteristics of SnO_2 nanofiber/TiO_2 nanoparticle composite for dye-sensitized solar cells. *AIP Advances*, *5*(6), 067134.

15. Fonash, S. J. Solar Cell Device Physics; Academic Press: New York, London, 1981.

16. Gregg, B. A.; Chen, S. G.; Ferrere, S. J. Phys. Chem. B 2003, 107, 3019.

17. Kozicki, M. N., Gopalan, C., Balakrishnan, M., Park, M., & Mitkova, M. (2004, November). Nonvolatile memory based on solid electrolytes. In *Proceedings. 2004 IEEE Computational Systems Bioinformatics Conference* (pp. 10–17). IEEE.

18. Shin, S., Kim, K., & Kang, S. M. (2010). Compact models for memristors based on charge-flux constitutive relationships. *IEEE Transactions on Computer-Aided Design of Integrated Circuits and Systems*, *29*(4), 590–598.

19. Jo, S. H., Chang, T., Ebong, I., Bhadviya, B. B., Mazumder, P., & Lu, W. (2010). Nanoscale memristor device as synapse in neuromorphic systems. *Nano Letters*, *10*(4), 1297–1301.

20. Chua, L. (1971). Memristor-the missing circuit element. *IEEE Transactions on Circuit Theory*, *18*(5), 507–519

21. Strukov, D. B., Snider, G. S., Stewart, D. R., & Williams, R. S. (2008). The missing memristor found. *Nature*, *453*(7191), 80–83.

22. Liu, B., Hu, M., Li, H., Mao, Z. H., Chen, Y., Huang, T., & Zhang, W. (2013, May). Digital-assisted noise-eliminating training for memristor crossbar-based analog neuromorphic computing engine. In *2013 50th ACM/EDAC/IEEE Design Automation Conference (DAC)* (pp. 1–6). IEEE.

23. Rajendran, J., Manem, H., Karri, R., & Rose, G. S. (2012). An energy-efficient memristive threshold logic circuit. *IEEE Transactions on Computers*, *61*(4), 474–487.

24. Shin, S., Kim, K., & Kang, S. M. (2011). Reconfigurable stateful NOR gate for large-scale logic-array integrations. *IEEE Transactions on Circuits and Systems II: Express Briefs*, *58*(7), 442–446.

25. Vourkas, I., & Sirakoulis, G. C. (2016). Emerging memristor-based logic circuit design approaches: A review. *IEEE Circuits and Systems Magazine*, *16*(3), 15–30.

26. Junsangsri, P., & Lombardi, F. (2012, May). A memristor-based TCAM (ternary content addressable memory) cell: Design and evaluation. In *Proceedings of the Great Lakes Symposium on VLSI* (pp. 311–314).

27. O'regan, B., & Grätzel, M. (1991). A low-cost, high-efficiency solar cell based on dye-sensitized colloidal TiO_2 films. *Nature*, *353*(6346), 737–740.

28. Arakawa, H., Yamaguchi, T., Sutou, T., Koishi, Y., Tobe, N., Matsumoto, D., & Nagai, T. (2010). Efficient dye-sensitized solar cell sub-modules. *Current Applied Physics*, *10*(2), S157–S160.

29. Grätzel, M. (2004). Conversion of sunlight to electric power by nanocrystalline dye-sensitized solar cells. *Journal of Photochemistry and Photobiology A: Chemistry*, *164*(1–3), 3–14.

30. Tulloch, G. E. (2004). Light and energy—dye solar cells for the 21st century. *Journal of Photochemistry and Photobiology A: Chemistry, 164*(1–3), 209–219.

31. Mathew, J., Ottavi, M., Yang, Y., & Pradhan, D. K. (2014, October). Using memristor state change behavior to identify faults in photovoltaic arrays. In *2014 IEEE International Symposium on Defect and Fault Tolerance in VLSI and Nanotechnology Systems (DFT)* (pp. 86–91). IEEE.

32. Giordano, F., Guidobaldi, A., Petrolati, E., Mastroianni, S., Brown, T. M., Reale, A., & Di Carlo, A. (2011, September). PSpice models for dye solar cells and modules. In *2011 Numerical Simulation of Optoelectronic Devices* (pp. 43–44). IEEE.

33. Giordano, F., Petrolati, E., Brown, T. M., Reale, A., & Di Carlo, A. (2011). Series-connection designs for dye solar cell modules. *IEEE Transactions on Electron Devices, 58*(8), 2759–2764.

34. Sastrawan, R., Renz, J., Prahl, C., Beier, J., Hinsch, A., & Kern, R. (2006). Interconnecting dye solar cells in modules—I–V characteristics under reverse bias. *Journal of Photochemistry and Photobiology A: Chemistry, 178*(1), 33–40.

35. Yang, Y., Mathew, J., Shafik, R. A., & Pradhan, D. K. (2013). Verilog-A based effective complementary resistive switch model for simulations and analysis. *IEEE Embedded Systems Letters, 6*(1), 12–15.

36. Rajendran, J., Manem, H., Karri, R., & Rose, G. S. (2012). An energy-efficient memristive threshold logic circuit. *IEEE Transactions on Computers, 61*(4), 474–487.

37. de Oliveira Reiter, R. D., Michels, L., Pinheiro, J. R., Reiter, R. A., Oliveira, S. V. G., & Péres, A. (2012, November). Comparative analysis of series and parallel photovoltaic arrays under partial shading conditions. In *2012 10th IEEE/IAS International Conference on Industry Applications* (pp. 1–5). IEEE.

7 Development of IoT-Based Data Acquisition System for Real-Time Monitoring of Solar PV System

Reetu Naudiyal and Sandeep Rawat
University Polytechnic, Uttaranchal University, Dehradun, India

Safia A. Kazmi
ZHCET, Aligarh Muslim University, Aligarh, India

Rupendra Kumar Pachauri
University of Petroleum and Energy Studies, Dehradun, India

CONTENTS

7.1 INTRODUCTION

The solar PV plant is an example of a non-conventional renewable energy source that has received technological maturity. In many applications where traditional electricity solutions become almost impossible to construct, its characteristics and working functions make this extremely interesting. Solar PV system operates over a long duration of time and generates performance data in terms of voltage, current, temperature, humidity, and irradiation of sunlight. In this modern civilization, electrical power plays an important role. For industries and human civilization, the rightful development of any nation will fully rely and depend on the accessibility and

123

availability of electrical energy. Due to technological development and considerable amount of urbanization, the world's energy consumption is increasing very fast. The traditional method for data collection from the solar PV system is manual, in which conventional instruments such as multimeters and other analog devices are used, which is a time-consuming process.

Due to changes with the time in surrounding conditions, accurate reading manually is challenging to obtain. To overcome this problem, automatic sensor based data acquisition system (DAS) is designed for the real world, which will provide error free and rapid response. In real time, the whole process of using DAS becomes more accurate compared with manual measurement; both are there for monitoring and analysis of the PV system's output.

7.2 LITERATURE REVIEW

Normally a solar PV system is characterized manually, which takes too much time and leads to error-based readings. For taking good performance of PV systems, DAS is designed and developed. In Ref. [1], the authors have established an experimental setup for data measurement techniques at an isolated site for a PV system to control its functioning with a microcontroller system. In Ref. [2], the authors have designed and implemented a DAS for hybrid energy systems for performance parameters. The developed system is a low-cost, sensitive, reliable, and easy system with programming capability features. In Ref. [3], the authors have discussed work on the photovoltaic (PV) characterization of conversion, which is based on a computer-based instrumentation. The designed system is used for PV module characterization in real-time meteorological test conditions. The DAS presents a good performance, compared with the cards on the market, and at a low cost. The authors of Ref. [4] discussed a new concept on the low-cost DAS for the renewable energy plant using a USB interface. This DAS works successfully on both Linux and windows operating systems and gives good results in terms of performance, reliability, sensitivity, and easy programming. In Ref. [5], the authors introduced a real-time online monitoring system for stand-alone PV applications where the online monitored data is further processed to provide a complete analysis of the PV system performance using a LAB-view-based graphical user interface (GUI). Furthermore, a web portal where an external user can see all monitored and derived parameters' daily evolution is also available. In Ref. [6], for the analysis of voltage and current produced by a solar–wind hybrid power system, the authors have designed data logger for reshaping processes, and then the result that comes from the data logger is used to reshape an inefficient and ineffective generating system. The authors of Ref. [7] designed and developed a novel data logger using the open-source electronics platform of monitoring PV systems, especially in remote areas. It was examined under various types of harsh environmental conditions during a 6-month period and the result was that it is reliable and monitored all the requested parameters with high accuracy. In Ref. [8], the authors have introduced a data logger named Nano-Logger, which is based on Arduino (open-source platform), featuring a customized circuit board, a real-time clock (RTC), and a micro SD card (memory purpose). The authors of Ref. [9], using the Arduino-Mega platform, built a low-cost system for thermal monitoring. The advantages of using the multiplex system is that there is elimination of thermocouple wires; installation of these will modify the fluid flows in the thermal reservoir. In Ref. [10], the authors designed a general data logger for a 240 W photovoltaic (PV) system that can store bulk data from input channels in large memory storage. This data logger justified reliability in monitoring the PV system and getting nearly ±0.5% accuracy by comparing the data obtained from the proposed data logger with the information obtained from the existing data logger on the market, DT80. They also built a simulation model for the acquisition of real-time solar panel data in LabVIEW. In a prototype-developed model, the authors have used two Arduino systems. First one Arduino system is used for the interfacing of PV system with the PC and the second one for data acquisition with the servomotor system. The variation of voltage and behavior of the PV system were recorded during different days and for different durations of time. The authors have also developed the test bench for the photovoltaic-thermal (PV-T) system. Without any interruption in the data acquisition, the system's reliability was verified during 6 months. The test bench uses simple, an open-source, economic, and robust system [12]. In Ref. [13], the authors have designed a low-cost DAS for monitoring a PV system's voltage, current, state of charge of the battery, and battery temperatures. Data sets are analyzed, and the overall performance can be observed and if the value will decrease which can lower the cost of DAS and allow for increased customization. For a stand-alone PV system, for continuously collecting and displaying the electrical output parameters, a low-cost data acquisition system was developed, which is based on Lab-VIEW. In Refs. [14,15], the authors have designed and tested a DAQ device to monitor the solar PV system performance parameters. To continuously monitor the performance parameters such as temperature, ambient temperature, voltage, current, and humidity of PV system, a prototype model was developed and stored in DAS with 10 ms time interval. In Ref. [16], for the $I–V$ and $P–V$ characteristics of PV panel processed under real-time conditions, the authors developed a virtual instrument with a low-cost Arduino acquisition board that can be obtained directly and plotted on an Excel spreadsheet without reprogramming the microcontroller. In Ref. [17], the authors proposed that, in Arduino, direct current (DC) motor and H-bridge motor driver circuit, and a wireless sensing controller can be utilized to change the tilt point of the PV panel, which follows the Sun, while the azimuth and the elevation angles were fixed around early afternoon. The system is very simple while providing good solar-tracking results and efficient power outputs. In Ref. [18], the authors developed an open-source and low-cost power monitoring system capable of monitoring different types of measurements, including loads and supplies such as solar PV systems. The power monitoring system can be fabricated using standard distributed manufacturing techniques, furthering its utility. A low-cost grid-connected PV system is developed for real-time monitoring of PV systems. The developed system has various features; it is inexpensive, hardware and software are easily available and accessible and do not require any highly specific skills. The system will provide all the considerable requirements and features in terms of accuracy and reliability [19]. In Ref. [20], the authors have introduced a stand-alone data logger device to measure the power characteristics of a solar panel in real-time with 3% of error in measurement, which indicates that the current sensor could measure the current with a similar result as compared to the calibrated standard lab instruments. In Ref. [21], for intelligent remote and real-time DAS monitoring, the authors have implemented an Internet of Things (IoT) that is based on data monitoring of the PV system. It was observed that the data is collected with 98.49% of accuracy and is able to send the data in graphical representation to a smartphone application with a mean transmission time of 52.34 seconds. In Ref. [22], the authors have designed a system to monitor solar photovoltaic system (SPV) parameters in real time using the IoT. For real-time monitoring of the PV panel of output voltage, power, current, and temperature, a complete application is developed on an android studio for mobile application. The authors of Ref. [23] designed an open-source internet of things for providing a cost-effective solution in real time that could collect and monitor the produced power and environmental conditions of solar stations in an intelligent manner. To decrease the cost of the measurement instruments, the designed solution is based as a laboratory prototype. When there is a mismatch of power quantity of solar power generation from the predefined set of standard values, the system always provides an alert to a remote user. The experimental test bench is increased by integrating alerts for any abnormality in power detections in PV power stations.

As every device is developed for an alternate, explicit situation, each contains lacks inside their structures when employed for different situations. The DAS to be structured in this paper is for the initial researcher. It is used to calculate the I–V and P–V characteristics of various PV panels. Because of its distant nature and cost contemplations, the DAS described in this paper employs micro-SD cards and IoT for its data storage and transfer of real-time results. This will permit the intercontinental transfer of the data and enable large data sets to be stored. A microcontroller and sensors are used to acquire PV system parameters such as voltage, currents, temperature, irradiation, and PV panel humidity.

In this chapter, the proposed design of DAS is based on economic aspect especially. With this design idea, the system considers the difficulty of a constant physical collection of data using micro-SD cards. The DAS can be integrated without high-level skill with the PV system to operate it in a real-time environment.

7.3 HARDWARE SYSTEM DESCRIPTION

The proposed hardware of DAS is used to calculate the I–V and P–V characteristics of PV solar panels using ATmega328p and other sensing components. The concept of this project is to develop DAS for PV panel characteristics. The proposed system can also save the parameter of the 20 W PV systems within a long-term time range. The circuit consists of a current sensor, voltage sensor, SD card module, DS3231 RTC module, and light intensity sensor (BH1750). The ATmega328p microcontroller that is compatible with Arduino IDE is used. The designed features are selected as Arduino IDE-based microcontroller, which is user-friendly and more easy to develop compare to other card modules. Data is stored on the SD Card as well as available directly in real time. The design of the hardware system is given in Figure 7.1.

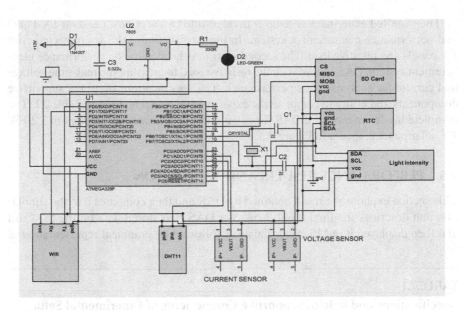

FIGURE 7.1 Circuit diagram of DAS.

7.4 MODELING OF DAS

The system is designed only for the experimental purpose, which shows the DAS compact dimensions and weight for the further development until there is a mass product. Developments of this project reduce the researcher's time and give characteristics of the different 20W PV panels. The developed system is shown in Figure 7.2.

FIGURE 7.2 Hardware model of DAS.

The installed experimental setup is divided into two sections, i.e. solar PV Panel and performance measurement system. In the first section, DAS is connected with PV panels and resistive load. The second section is headed by the performance measurement system. DAS is used with the resistive load to measure the real-time voltage and current for analysis of the performance. The specifications of all the supportive components and utility to comprise the experimental set up is listed in Table 7.1. The complete hardware model of DAS for calculation of different parameters in a 20 W PV solar panel is shown in Figure 7.3.

7.5 PERFORMANCE RESULTS AND VALIDATION

This section explains the result obtained by DAS and then compares it to the simulation and describes the final result. Low-cost DAS first stored data on the SD Card and then displayed it in MS excel. This was followed by graphical representation of

TABLE.7.1

Specifications and Role of Supportive Components of Experimental Setup

Section	Components	Specifications	Role/Function
PV solar Panels	Usha PV module (20 W)	O. C. voltage: 21.997 V S. C. current: 1.2586 A I_{mpp}: 1.12A, V_{mpp}: 18 V Cell technology: Poly-Si Dimension (mm): 356×490×25 Manf.: USHA SHRIRAM Technologies (Model NO: US 20/12 V)	PV module is used to draw the short circuit current to the load
	Solar Spark PV module (20 W)	O. C. voltage: 22.58 V S. C. current: 1.19 A I_{mpp}: 1.08 A, V_{mpp}: 18.82 V Cell technology: Poly-Si Dimension (mm): 356×490×25 Manf.: SOLAR SPARK SHRIRAM Technologies (Module type: SS 20-18-P)	PV panel is used to draw the short circuit current to the load
	MicroSun PV module (20 W)	O. C. voltage: 21 V S. C. current: 1.36 A I_{mpp}: 1.16 A, V_{mpp}: 17.1 V Cell technology: Poly-Si Dimension (mm): 650×300×450 Manf.: MIRCOSUN SOLAR (Model NO: 1220/ SL.NO.: 18176532)	PV panel is used to draw the short circuit current to the load

(Continued)

TABLE.7.1 (*Continued*)

Specifications and Role of Supportive Components of Experimental Setup

Section	Components	Specifications	Role/Function
Performance measurement system	Data Acquisition System	Number of Current Sensor: 1 (0–5 A) Number of Voltage sensor: 1(0–25 V) Number of Light intensity sensor : 1 (0–65535 l × unit) Number of SD-Card module: 1 (20×28 m) Number of RTC Module: 1 Number of Humidity and temperature sensor: 1 Breadboard: 1 (9×6×0.1 cm)	To measure the following panel parameters: Voltage, Current, Temperature, Humidity, and irradiation of halogen lamp with time and date of the measurement.
	Decade resistive load	Number of resistive load:2 Range: 0.1–250 Ω	Variable load (decade resistive box)
	Artificial solar lamp	Total number of lamps-2 Light intensity 50–650 W/m^2	The solar lamp system for uniform light intensity on PV system

FIGURE 7.3 Complete hardware model of DAS.

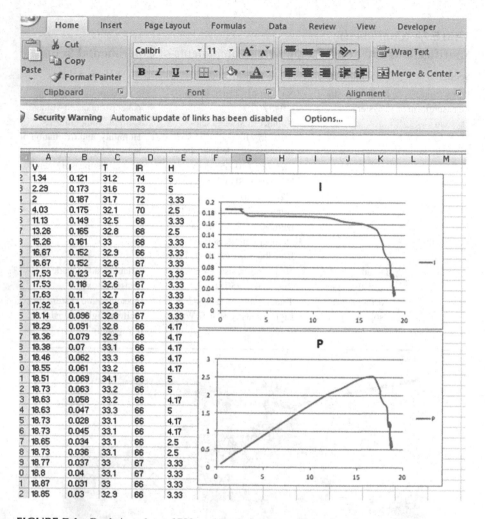

FIGURE 7.4 Real-time data of PV module during experimental study.

the results obtained by the DAS of the PV panel. The collected real-time datasheet is shown in Figure 7.4.

For performance validation, MATLAB®/Simulink modeling of the PV system is carried out. For these purposes, three different PV systems (20 W) are considered to characterize the experimental and simulation study. In the results and discussion section, all the obtained results in terms of I–V characteristics are compared and shown in Figure 7.5.

These I–V characteristics of three different PV modules panels are shown under three different irradiation levels, 650 W/m^2, 360 W/m^2, and 170W/m^2. In the P–V characteristics, the dark lines show the performance for low-cost DAS and the dotted line shows the performance under the simulation study. Under similar environmental conditions, P–V characteristics are given in Figure 7.6.

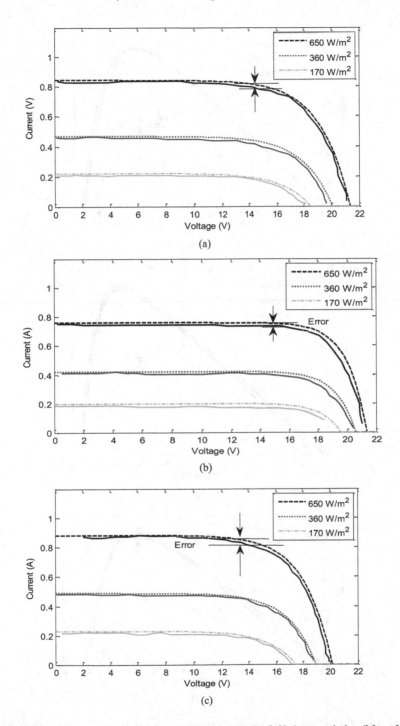

FIGURE 7.5 *I–V* characteristics of solar PV systems. (a) *I–V* characteristics (Manuf. Usha 20 W PV system), (b) *I–V* characteristics (Manuf. Spark Solar 20 W PV system), (c) *I–V* characteristics (Manuf. MicroSun 20 W PV system).

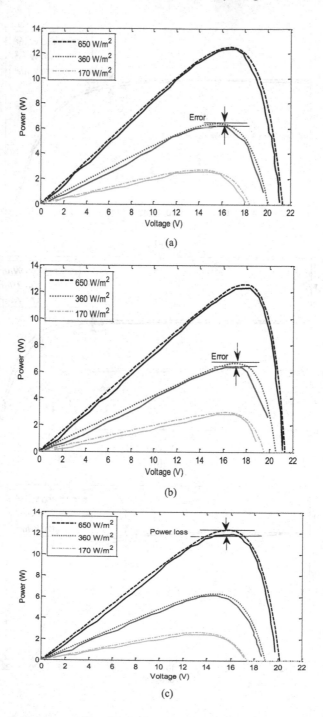

FIGURE 7.6 *P–V* characteristics of solar PV systems. (a) *P–V* characteristics (Manuf. Usha 20 W PV system), (b) *P–V* characteristics (Manuf. Spark Solar 20 W PV system), (c) *P–V* characteristics (Manuf. MicroSun 20 W PV system).

The results obtained from *I–V* and *P–V* characterizations of all three types PV systems are summarized in Tables 7.2 and 7.3.

By comparing the values in Tables 7.2 and 7.3, power losses can be calculated using Equation (7.1) and are summarized in Table 7.4:

$$\text{Error}\% \frac{\text{estimated value} - \text{actual value}}{\text{actual value}} \times 100 \qquad (7.1)$$

The cost analysis of developed system is given in Table 7.5.

7.5.1 DISCUSSION ON OBSERVATIONS

An attempt has been made to present the best and low-cost DAS to produce a modest means of collecting PV panel data. Graphical representation of the data in MATLAB/ simulation was done and compared with the DAS result, which proves system accuracy. This mechanism of DAS is free of observational and environmental error and is used in a wide range of irradiation systems. The panels under consideration for observation were kept under halogen lamps due to which the short-circuit current (I_{sc}) at no-load and the observations were recorded on the DAS simultaneously. It was observed that with the increase in load, is decreasing and voltage increase, which is analyzcd in DAS in real-time. Along with the voltage and current, the DAS records the temperature and humidity of the experimental setup, which affects the DAS observed; the observations were performed at three different levels of irradiation in the same manner; after that, the MATLAB graphs were plotted and compared with the simulation results, which shows the accuracy of DAS. The working performance of the DAS and the results and comparison with the simulation proved that the DAS is the best means of collecting and recording the PV panel and helpful design and develop the PV panel in different Irradiation levels and at different power levels.

Advantages

a. A Flexible mechanism is developed that can dynamically provide the PV panel parameters and specification in real time with the calculation of

---

TABLE 7.2

Performance Parameters of MATLAB/Simulink Study

Panel	Usha PV Panel			Solar Spark PV Panel			Micro Sun PV Panel		
Specification	650 W/m²	360 W/m²	170 W/m²	650 W/m²	360 W/m²	170 W/m²	650 W/m²	360 W/m²	170 W/m²
V_{oc}	21.3	19.74	18.44	21.3	20.43	19.26	19.91	18.81	17.13
I_{sc}	0.844	0.46	0.22	0.76	0.421	0.198	0.88	0.48	0.23
V_{mx}	16.77	16.03	14.66	17.57	17.13	16.43	16.03	15.05	13.49
I_{mx}	0.74	0.39	0.18	0.71	0.38	0.18	0.76	0.41	0.19
P_{mx}	12.5	6.40	2.69	12.59	6.67	2.96	12.31	6.28	2.66
FF	0.69	0.69	0.65	0.77	0.76	0.77	0.69	0.68	0.65

TABLE.7.3
Performance Parameters of Experimental Study

Panel	Usha PV Panel			Solar Spark PV Panel			Micro Sun PV Panel		
Specification	650 W/m^2	360 W/m^2	170 W/m^2	650 W/m^2	360 W/m^2	170 W/m^2	650 W/m^2	360 W/m^2	170 W/m^2
V_{oc}	21.07	16.58	18.03	21.01	20.43	18.45	19.91	18.75	17.11
I_{sc}	0.836	0.459	0.207	0.75	0.42	0.18	0.87	0.48	0.21
V_{mx}	16.71	15.07	14.89	18.27	17.56	16.22	16.23	14.82	14.57
I_{mx}	0.73	0.40	0.17	0.67	0.36	0.17	0.73	0.41	0.16
P_{mx}	12.36	6.117	2.586	12.3	6.36	2.84	11.91	6.15	2.43
FF	0.69	0.79	0.67	0.77	0.76	0.73	0.68	0.67	0.64

TABLE.7.4
Calculated Power Loss and Error Percentage

	Usha PV Panel		P_{loss} of Solar Spark PV Panel		P_{loss} of MicroSun PV Panel	
Irradiation (W/m^2)	Power Loss (W)	Error (%)	Power Loss (W)	Error (%)	Power Loss (W)	Error (%)
650	0.14	1.13	0.29	2.3	0.4	3.3
360	0.283	4.6	0.31	4.8	0.13	2.1
170	0.104	4	0.12	4	0.23	9.4

TABLE.7.5
Price List of DAS Component

Name of Component	Specification	Price (INR)
Voltage sensor module	0–25 V	300
Current sensor ACS712 module	5 A	200
Light intensity sensor BH1750 module	65535 1×units	150
SD card module	-	100
DS3231 RTC module	-	150
DHT11 temperature and humidity sensor module	-	300
Arduino ATMega328P	-	300
	Total cost:	3500

meteorological effects on panels and restore the data any time. For performance assessment purposes, the data should be collected regarding the installed system performance. DAS are broadly used in renewable energy source (RES) applications in order to collect data.

b. The DAS's reliability is because no personal installation schemes and economic considerations are required, and this technique can predict the performance of the PV panel and prove beneficial to foretell about its characterization before their installation. Also, a better understanding can improve the reliability of the solar panel.

c. The algorithm is flexible and allows the user to specify the number of modules integrated into the array. Since faster iterative techniques have been used, the modeling approach is instrumental in characterizing the array design under consideration quickly. With slight modifications, it is possible to study different architectures for performance prediction under random shading patterns.

d. Since switching occurs between the modules that exist within the array, no additional bank of solar cells needs to be added. Thus, the overhead of extra hardware is reduced, making the system less costly and more efficient.

e. Considerable improvement is observed in the fill factor for systems employed with switching over the ones with no switching.

Modeling of the solar modules with bypass has been integrated with switching the modules in a bigger array system. The systems devoid of switching mechanism and bypass diodes have been compared with the cumulative effects of both system. There are several situations that have been documented with performance characteristics, such as I–V and P–V curves.

7.6 CONCLUSION

The salient points are as follows:

- The obtained results were satisfactory, and the developed DAS provided better accuracy, compared with commercial systems, and reliability.
- The performance parameters are analyzed through I–V and P–V curves.
- A performance validation of the results obtained through the experiment and MATLAB study was carried out for extensive analysis.

REFERENCES

1. M. Benghanem, "Low Cost Management for Photovoltaic Systems in Isolated Site with New IV Characterization Model Proposed," *Energy Conversion and Management*, vol. 50, no. 3, pp. 748–755, 2009.
2. M. Demirtas, I. Sefa, E. Irmak, and I. Colak, "Low-Cost and High Sensitive Microcontroller Based Data Acquisition System for Renewable Energy Sources," in *Proceedings IEEE Conference on Power Electronics, Electrical Drives, Automation and Motion*, 11–13 June 2008 at Italy, pp. 196–199, 2008.

3. H. Belmili, S. M. A. Cheikh, M. Haddadi, and C. Larbes, "Design and Development of a Data Acquisition System for Photovoltaic Modules Characterization," *Renewable Energy*, vol. 35, no. 7, pp. 1484–1492, 2010.

4. S. C. S. Juca, P. C. M. Carvalho, and F. T. Brito, "A Low Cost Concept for Data Acquisition Systems Applied to Decentralized Renewable Energy Plants," *Sensors*, vol. 11, pp. 743–756, 2011.

5. M. Torres, "Online Monitoring System for Stand-Alone Photovoltaic Applications — Analysis of System Performance from Monitored Data," *Journal of Solar Energy Engineering*, vol. 134, no. 3, pp. 1–8, 2017.

6. M. Ikhsan, A. Purwadi, N. Hariyanto, N. Heryana, and Y. Haroen, "Study of Renewable Energy Sources Capacity and Loading Using Data Logger for Sizing of Solar-Wind Hybrid Power System," *Procedia Technology*, vol. 11, pp. 1048–1053, 2013.

7. M. Fuentes, M. Vivar, J. M. Burgos, J. Aguilera, and J. A. Vacas, "Design of an Accurate, Low-Cost Autonomous Data Logger for PV System Monitoring Using Arduino That Complies with IEC Standards," *Solar Energy Materials & Solar Cells*, vol. 130, pp. 529–543, 2014.

8. M. Gandra, R. Seabra, and F. P. Lima, "Oceanography Methods: A Low-Cost, Versatile Data Logging System for Ecological Applications," *Association of the Sciences of Limnology and Oceanography*, vol. 13, no. 3, pp. 115–126, 2015.

9. E. Avallone, D. Garcia, C. Alcides, P. Vicente, and L. Scalon, "Electronic Multiplex System Using the Arduino Platform to Control and Record the Data of the Temperatures Profiles in Heat Storage Tank for Solar Collector," *International Journal of Energy Environment Engineering*, vol. 7, no. 4, pp. 391–398, 2016.

10. N. N. Mahzan, A. M. Omar, L. Rimon, S. Z. M. Noor, and M. Z. Rosselan, "Design and Development of an Arduino Based Data Logger for Photovoltaic Monitoring System Design and Development of an Arduino Based Data Logger for Photovoltaic Monitoring System," *International Journal of Instrumentation of Control System*, vol. 7, no. 3, pp. 15–25, 2017.

11. S. Mandal, and D. Singh, "Real Time Data Acquisition of Solar Panel Using Arduino and Further Recording Voltage of the Solar Panel," *International Journal of Instrumentation of Control System*, vol. 7, no. 3, pp. 15–25, 2017.

12. C. Ulloa, J. M. Nunez, A. Suarez, and C. Lin, "Design and Development of a PV-T Test Bench Based on Arduino," *Energy Procedia*, vol. 141, pp. 71–75, 2017.

13. S. Fanourakis, K. Wang, P. M. Carthy, and L. Jiao, "Low-Cost Data Acquisition Systems for Photovoltaic System Monitoring and Usage Statistics," In *Proceedings IOP Conference Series-Earth and Environment*, pp. 719–728, 2017.

14. H. Rezk, I. Tyukhov, M. A. Dhaifallah, and A. Tikhonov, "Performance of Data Acquisition System for Monitoring PV System Parameters," *Measurement*, vol. 104, pp. 204–211, 2017.

15. N. Sugiartha, I. M. Sugina, and I. D. G. T. Putra, "Development of an Arduino-based Data Acquisition Device for Monitoring Solar PV System Parameters," In *Proceedings International Conference on Science and Technology*, pp. 995–999, 2018.

16. A. E. Hammoumi, S. Motahhir, A. Chalh, A. El Ghzizal, and A. Derouich, "Low Cost Virtual Instrumentation of PV Panel Characteristics Using Excel and Arduino in Comparison with Traditional Instrumentation," *Renewables Wind, Water and Solar*, vol. 5, pp. 1–16, 2018.

17. S. Kyi, and A. Taparugssanagorn, "Wireless Sensing for a Solar Power System," *Digital Communication Networks*, vol. 6, pp. 51–57, 2020.

18. S. Oberloier and J. M. Pearce, "Open Source Low-Cost Power Monitoring System," *HardwareX*, vol. 4, pp. 1–21, 2018.

19. N. Erraissi, M. Raoufi, N. Aarich, M. Akhsassi, and A. Bennouna, "Implementation of a Low-cost Data Acquisition System for "PROPRE.MA" Project," *Measurement*, vol. 117, pp. 21–40, 2018.
20. M. S. Hadi, A. N. Afandi, A. P. Wibawa, A. S. Ahmar, and K. H. Saputra, "Stand-Alone Data Logger for Solar Panel Energy System with RTC and SD Card", In *Proceedings International Conference on Statistics, Mathematics, Teaching, and Research*, pp. 1–9, 2018.
21. W. Priharti, A. F. K. Rosmawati, and I. P. D. Wibawa, "IoT Based Photovoltaic Monitoring System Application," In *Proceedings International Conference on Engineering, Technology and Innovative Researches, IOP Publishing, Journal of Physics: Conference Series*, pp. 1–10, 2019.
22. S. Sarswat, I. Yadav, and S. K. Maurya, "Real Time Monitoring of Solar PV Parameter Using IoT," *International Journal of Innovative Technology and Exploring Engineering*, vol. 9, pp. 267–271, 2019.
23. Y. Cheddadi, H. Cheddadi, F. Cheddadi, F. Errahimi, and N. Essbai, "Design and Implementation of an Intelligent Low-Cost IoT Solution for Energy Monitoring of Photovoltaic Stations," *SN Applied Sciences*, vol. 2, pp. 1–11, 2020.

8 Marine Photovoltaics – An IoT-Integrated Approach to Enhance Efficiency

R. Raajiv Menon and Jitendra Kumar Pandey
University of Petroleum and Energy
Studies, Dehradun, India

R. Vijaya Kumar
Indian Institute of Technology IIT(M), Chennai, India

CONTENTS

8.1 INTRODUCTION

The rapidly depleting fossil fuel reserves and increasing volatility in the Middle East oil-rich nations has forced world nations to recourse its energy strategy towards dependency on coal and other fossil fuels (Menon, 2014). The other energy alternative such as nuclear and unconventional energy sources, viz., shale oil, coal bed methane (CBM), and natural gas hydrates, however pose an interesting alternate option but are often coupled with higher investment price and imminent danger to mankind as well as the environment. The concept of renewable energy has slowly gained pace over the last 30–40 years due to the sustained efforts of developed nations and other environmental protection agencies to curb the Greenhouse Gas (GHG) emissions. Renewable energies, viz., solar, wind, geothermal, wave, bioenergy, and hydro, have gained popularity in the world as an alternate source of energy and has come of age to be acceptable as an independent energy sources powering the day-to-day needs of the nation. The Asia-Pacific region accounts for more than 85% of the total consumption of conventional fossil fuels ("Asia and Pacific," 2020).

The ozone layer depletion ("Chlorofluorocarbons and Ozone Depletion – American Chemical Society," n.d.; Thangavel, 2018) due to GHG emissions has caused major catastrophic phenomena including global warming, melting of Arctic and Antarctic ice glaciers, and depletion of water reservoirs, thus necessitating mankind to look for a greener source of energy consumption. These renewable energies will not only reduce the dependency on fossil fuels but will also aid in repair of the environment. Several international agencies and programs, International Maritime Organisation (IMO) ("International Maritime Organization," n.d.), Intergovernmental Panel on Climate Change (IPCC), and United Nations Environment Protection (UNEP) ("UNEP – UN Environment Programme," n.d.), impose severe restrictions on the use of fossil fuels. The UNEP has chalked out many impressive programs, viz., Montreal protocol (protection of ozone layer; Zerefos, Contopoulos, & Gregory, 2009) and solar loan program (with attractive financing has had successful implementation and exploitation of renewable source in India, Mexico, Indonesia, Tunisia, and Morocco; "UNEP's India Solar Loan Programme Wins Energy Globe," 2007).

With the given impetus for the need of renewable energy, intergovernmental agencies like the International Renewable Energy Agency (IRENA) has been flagbearers for promotion of various kinds of sustainable renewable energy with the cooperation of over 180 countries. IRENA also undertakes mapping of renewable energy projects, Planning and formulation of Renewable Energy map (Remap) for 2030 ("About IRENA," n.d.). Individual countries have provided incentives to green transportation vehicles utilizing hydrogen and electricity as powering sources so as to curb the dependency on fossil fuels. Rapid development and research in the field of IT has paved way for amalgamation of all sensors, embedded systems, control equipment, and automation into a single entity known as Internet of Things (IoT). IoT stands to invigorate the world energy scenario with its greater ability to function without a man or machine interface. IoT has found compatibility in every sector, including civil and military domain, encompassing a variety of fields, viz., transportation, health, energy, defense, agriculture, and environment.

The renewable energy sector has shown tremendous potential towards adaptability in the past 50 years; with evolving concepts like microgrids and hydrogen-powered DC infrastructure, these technologies with effective IT support will be pillars of strength in the energy scenarios in the future. The unique advantages offered by these microgrids often finds employability as major powering source in remote isolated terrains and on seas. The dependency of powering an island nations and its fossil fuel-dependent infrastructure can be reworked to incorporate renewable clean, green energy sources. With its widespread abundance and availability, solar power emerges as a strong contender among other renewables with its widespread prevalence on land as well as on the water fronts. The solar platforms not only aid towards generation of power but also helps in improving the efficacy of irrigation in the agriculture sector and prevents evaporation losses in water bodies. IRENA reports that an estimated 580 GW of solar energy systems were installed by 2019 (Renewable Energy Agency International IRENA, 2019). Asian regions have contributed to 60% of the overall solar installation between 2010 and 2019. With decreasing costs and improved efficiency towards manufacturing of sustainable solar panels, the renewable energy source has promoted a competitive market. Operational and maintenance costs have steadily decreased over the period of time due to real-time analysis facilitated by big data and other Artificial Intelligence (AI) programs. This predictive analysis software enabled to cut down the system downtime and increased the productivity output of the overall system (Lee, Cao, & Ng, 2017).

8.2 LAWS AND REGULATIONS IN THE MARITIME SECTOR

International organizations, viz., European Commission (EC) and International Maritime Organisation (IMO), have promulgated various emission control standards towards reduction of harmful GHGs emanating from ships. The annual estimate of fuel oil consumption in the shipping sector is estimated to be 87 billion gallons approximately and is expected to double by the year 2030. Marine sector, being one of the largest consumers of fuel oil, contributes to the biggest noxious emissions in this sector. The black carbon in artic region is attributed to the SO_x emissions from the shipping sector. These emissions cause severe deterioration of the environment and cause severe health issues to living beings across the world. The regulation of emission standards is aggressively monitored by IMO. The most prominent among them is to reduce the sulfur content in the emission from 3.5% to 0.5% by 2020. Further, it has also promulgated the Emission Control Areas (ECAs) in developed countries like the United States and Europe, which restricts sulfur emissions to 0.1% for coastal vessels operating close to ECAs. These regulations will entail maritime operators to substitute their present usage of Heavy Fuel Oil (HFO) to distillate oils, viz., Marine Diesel Oil (MDO) or Marine Gas Oil (MGO). These distillate oils have relatively lesser sulfur content as compared to HFO. Adaptation to the 2020 standards will necessitate the following:

a. Installation of scrubber equipment – Exhaust cleansing systems as an immediate measure till suitable studies/modifications have been made onboard towards adaptation of distillate fuels in lieu of the currently used HFO. These scrubbers are fairly cheaper than the distillate fuel usage and, hence, have greater and immediate influence onboard older and relatively new marine platforms.

b. Utilization of alternate fuel – use of methanol and its blended variants, biodiesels (Demirbas, 2007; Stead, Wadud, Nash, & Li, 2019; Yusoff et al., 2020; Zailani, Iranmanesh, Hyun, & Ali, 2019), hydrogen, and liquified petroleum gas (LPG) are considered the current best alternatives to fossil fuels. Several research studies have been undertaken towards adaptation of these alternative fuels onboard marine platforms. However, these alternatives are presently constrained by various factors including a narrower market. Storage and production options, requirement of modified machinery to suit alternative fuels, Operational and Capital expenditure surrounding the switchover to alternative fuel.

c. The current IMO regulations as of 2020 requires all vessels to switch over to distillate fuels (MDO, MGO) from HFO (Colcomb, Rymell, & Lewis, 2006; Comer, 2019). All measures are to be undertaken in order to keep the sulfur emissions in check.

8.3 SOLAR ENERGY POTENTIAL

Solar energy has tremendous potential to power the entire earth's energy requirements. Technological advancements along with exponential research in this field provide affordable clean energy. The radiated energy of the sun is utilized for a variety of applications that provides electricity, heat water, and store energy. There are various types of solar energy utilities, viz., photovoltaics (PV), solar thermal systems (STS), and concentrating solar power (CSP), which contribute to the overall utilization of the solar energy potential. More than 120 world nations have pledged their unprecedented support towards implementation of solar energy sources. Owing to the increased incentives provided by countries, solar PVs have been able to provide electricity to remote locations, thus reducing the electrification gap within the country. The statistical analysis report of the British Petroleum (BP) company suggests that oil remains the primary source of energy in many developing nations. The statistics of proven oil reserves as of 2019 is shown in Figure 8.1.

Among the renewable energy sources, solar energy has been the most productive, contributing to the majority of power generation across the world. The renewable energy consumption as per the region vis-à-vis the source is shown in Figure 8.2.

The solar renewable era has progressed ahead in multidimension with introduction of newer concepts such as floating solar energy farms. These farms have

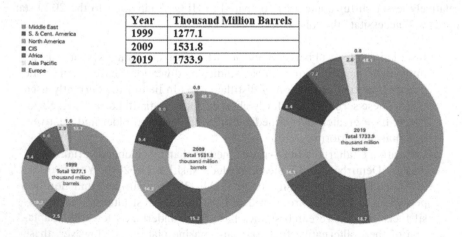

Year	Thousand Million Barrels
1999	1277.1
2009	1531.8
2019	1733.9

FIGURE 8.1 Distribution of region-wise proven oil reserves (BP, 2020).

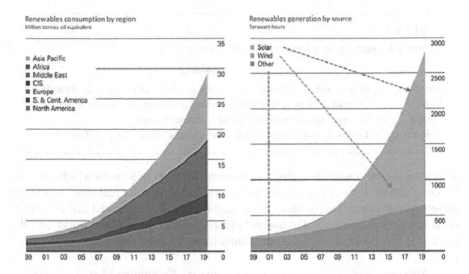

FIGURE 8.2   Renewable energy consumption as well as source distribution region wise (BP, 2020).

widespread acceptance and have been adapted over various countries because of its distinct advantages.

8.4   INTERNET OF THINGS (IoT) AND THE OCEAN

It is considered an amalgamation of information gathered from every possible sensor with unique identifiers transmitted back and forth over internet. These data are traversed into four layers, which form the backbone of the IoT architecture (Fersi, 2015; Ray, 2018; Sethi & Sarangi, 2017) as shown in Figure 8.3. The ground area

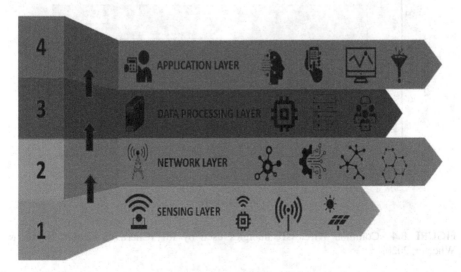

FIGURE 8.3   Internet of Things (IoT) architecture.

TABLE 8.1

Sensors Employed in a Floating Solar Farm

Sensor Type	Remarks
Position	Used to monitor the position of the farm/panel and GPS for accurate location
Light	To measure the amount of incident radiation and thus calculate the power output
Motion	To control the tilt angle of the panel to maximize output
Temperature	To Identify uniformity of temperature in panels/identify hotspots if any

incorporating the renewable energy is connected to a set of sensors which, in turn, are connected to IoT devices. These sensors efficiently transmit real-time data to the servers, which in turn enable smart processing of data. It is estimated that around 30 billion devices are connected to internet utilizing the IoT around the world, and the number is expected to increase exponentially in the forthcoming years.

a. Sensing Layer: The on-ground layer consists of a number of sensors, viz., position, temperature, light, motion, and pressure, that provide real-time inputs to the connected IoT devices which, in turn, transmits data to the network layer through the gateways. For example, the employability of sensors in a floating solar farm is shown in Table 8.1.

b. Network Layer: This layer acts as the communication medium between the sensors and the remote terminal. IoT employs a variety of wireless communication methodologies, viz., LoRa (Han et al., 2020; Zourmand, Kun Hing, Wai Hung, & Abdulrehman, 2019), Zigbee, LTE-M, Satellite, NB-IoT (shown in Figure 8.4), based on the customer's requirement and investment

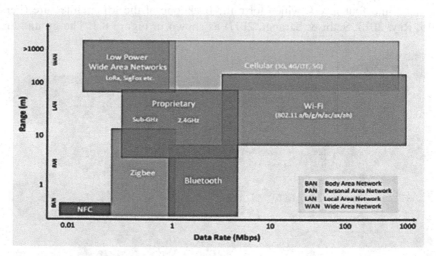

FIGURE 8.4 Common WirelessTechnolgies used by IoT (Cheruvu, Kumar, Smith, & Wheeler, 2020).

cost of the project. Naval IoT projects often utilize Satellite-based IoT for communication due to the wider range (>1500 km) and LoRa-Long Range low power wireless RF technology (1–10 km). Satellite technology is predominantly used where IoT devices are deployed in rural and remote areas (oceanic environment).

c. Data Processing Layer: This layer serves to process the field data collected from the sensing layer and further analyzes the data to provide a concrete solution based on the results. A very few IoT applications have options to store the analyzed data from previous data and employs them for achieving predictive reports based on previous history. The realized data are further sent to the application layer for visual appreciation and customer usage.

d. Application Layer: The processed data from the data processing layer are specifically processed and analyzed as per the customer's need and delineated protocols. Prognostic maintenance routines and system-initiated maintenance schedule can be initiated from the application layer (Saritha & Sarasvathi, 2019; Yassein, Shatnawi, & Al-Zoubi, 2016).

Wireless sensor networks (WSNs) (Bensaleh, Saida, Kacem, & Abid, 2020; Carlos-Mancilla, López-Mellado, & Siller, 2016) for the past few decades have been considered as a subset to the IoT technology primarily due to the widespread implementation of these devices in all fields, viz., transportation, smart cities, healthcare, and agriculture (Xu, Shi, Sun, & Shen, 2019). Due to the tremendous advantage of connectivity, it offers these devices that can be readily deployed in the marine environment for the purpose of monitoring and data collection in a marine environment. The traditional methods of research and data collection employed by hydrographic survey and oceanographic and scientific research vessels are extremely expensive and time bound when compared to the unmanned IoT-enabled vessels. The marine environment is often harsh and, hence, requires robust reliable systems in order to maximize the output and minimize the system downtime of these equipment. The vessels, viz., boats and buoys deployed in ocean, will contain a hybridized combination of solar and wind energies as their renewable energy source and will be integrated to the onboard battery monitoring system, which in turn shall govern the health status of the storage batteries. The maintenance team from a remote location can monitor the energy consumption and generation rate and can modulate the speed according to the ambient conditions of the ocean, for example, the vessel may be ordered to propel at a lower speed during night in order to conserve the battery charge. The IoT-deployed marine devices should possess the following:

a. Corrosion resistance
b. Stable connectivity
c. Efficient renewable and storage option
d. Robust construction
e. Hardware and software implementation to reduce system downtime

8.5 FLOATING PHOTOVOLTAICS

Though solar energy has been widely accepted as a cleaner energy source across the world, it has its own drawbacks, including massive land occupancy rate to match to the same energy equivalence produced by a nuclear or fossil fuel-dependent source. Second is the need for an efficient energy storage setup so as to cater during dark hours and seasonal weather. In order to tackle the land occupancy, floating solar farms have achieved much needed success; however, unlike any other renewable energy source, they require a dependable energy storage setup for smooth functioning. A variety of floating solar farms (Choi, 2014; S. H. Kim, Yoon, & Choi, 2017; S. M. Kim, Oh, & Park, 2019; Nebey, Taye, & Workineh, 2020) have swung into action for energy generation, viz., fixed floating solar farms, track-enabled floating solar farms, and submerged floating farms, depending on the regional conditions. The major advantages levied by these floating farms include

a. Decluttering of land occupancy – can be installed over any water surface and does not conflict with agricultural or any prospective land masses.
b. Prevention of evaporative loss from water bodies – due to larger area being covered. The evaporative losses during hot weather is greatly reduced by the presence of these floating farms.
c. The submerged floating solar farms reduces the requirement of cooling systems and due to the presence of water body and has shown a higher efficiency rate.
d. The cost of installation and dismantling of these floating farms is less and the processes are swift, efficient, and reversible as compared to the installations on land.
e. Due to the greater flexibility in movement provided by the floatation buoys, these can be maneuvered to increase the incident angle and obviate the requirement of complex tracking systems.
f. Floating solar farms offshore can be coupled with other renewable energy systems, viz., wind, wave, and ocean, to form complex hybrid renewable sources that can be further coupled to shore energy storage facility.

These floating energy platforms will coexist with nature and environment and often do not affect the ergonomics of the location.

The maintenance issues experienced in floating energy farms are often obviated by the IOT technology. The IOT often utilizes a real-time Supervisory Control And Data Acquisition (SCADA) network (Baker et al., 2020; Hunzinger, 2017; Sajid, Abbas, & Saleem, 2016; Shahzad, Kim, & Elgamoudi, 2017) to monitor the power generation and equipment condition of the floating farm ("Can Floating Solar Plants Break Free from the Trappings of High Maintenance?," n.d.). These systems are configured to undertake routine maintenance checks and initiate built-in self-checks whenever a malfunction occurs ("What Is SCADA? Supervisory Control and Data Acquisition," n.d.). The system, upon sensing a malfunction, generates an alarm signal in addition to initiation of maintenance check. The process can be remotely monitored by the customer or by the intermediate maintenance team. These IOT-based systems are put into use in order to undertake smart decisions with supported data analysis and, thus, mitigate system downtime. IOT-based solar farms have shown results in saving significant amount of time and money.

These floating farms, because of the close proximity of water, employ advanced predictive maintenance algorithms so that the maintenance team can undertake intermediate routine maintenance and prevent any major damage to the equipment. IOT sensors can identify any irregular hotspots on panel as well as rise in temperature due to humidity. Implementation of IOT technology in a floating solar farm will enable to optimize the power generation transmission and distribution in an efficient manner. These systems, in addition, will facilitate utilization of the generated electricity by having a positive control over the energy trading through implementation of block chain methodologies (Andoni et al., 2019; Edeland & Mörk, 2018; International Renewable Energy Agency – IRENA, 2019; Oh, Kim, Park, Roh, & Lee, 2017) (Table 8.2; Figure 8.5).

TABLE 8.2
Megawatt Power Floating Solar Farms Around the World

Plant Name	Capacity in MW	Location	No of Installed Panels	References
Anhui CECEP	70	Anhui Bengbu, Anhui, China	1,94,731	("ANHUI CECEP: 70,005 kWp – Ciel et Terre International," n.d.)
Sungrove Huainan Solar Farm	40	Anhui Province, Huainan city, China	1,60,000	("Huainan: Largest Floating Solar Farm in the World I Planète Énergies," n.d.)
Anhui GCL floating Solar Farm	32.69	Huaibei, China	1,16,736	("ANHUI GCL: 32,686 kWp – Ciel et Terre International," n.d.)
Piolenc Floating Solar Farm	17	Piolenc, France	47,000	("O'MEGA1: first floating solar power plant in France I Bouygues Energies & Services," n.d.)
Sekdoorn Project	14.5	Zwolle, Netherlands	40,000	("BayWa r.e. builds Netherland's biggest Floating Solar Farm – BayWa r.e.," n.d.) ("Project Briefing: New floating solar concepts at BayWa r.e.'s Sekdoorn I PV Tech," n.d.)
Yamakura Floating Solar Farm	13.7	Ichihara, Chiba Prefecture, Japan	50,000	("Yamakura Dam Welcomed Floating Solar Panels on its Surface – Ciel et Terre International," n.d.)

(Continued)

TABLE 8.2 (*Continued*)
Megawatt Power Floating Solar Farms Around the World

Plant Name	Capacity in MW	Location	No of Installed Panels	References
Pei County	9.98	Jiangsu, China	42,240	("PEI COUNTY: 9,982 kWp – Ciel et Terre International," n.d.)
Agongdian Floating Solar Farm	S 9.9	Kaohsiung, Taiwan	34,013	("AGONGDIAN: 9,994 kWp – Ciel et Terre International," n.d.)
Umenoki Floating Solar Farm	7.5	Higashimatsuyama city, Kanto, Japan	27,456	("UMENOKI: 7,550 kWp – Ciel et Terre International," n.d.)
GCL Jining Floating Solar farm	6.78	Jining, Shandong, China	24,640	("GCL JINING: 6,776 kWp – Ciel et Terre International," n.d.)
Queen Elizabeth II Reservoir Floating Solar Farm	6.3	London, UK	23,000	("World's biggest floating solar farm powers up outside London I Environment I The Guardian," n.d.)
Sugu 1 Floating Solar Farm	4	Tainan, Taiwan	13,410	("SUGU 1: 4,023 kWp – Ciel et Terre International," n.d.)
Sayreville Floating Solar Farm	4.4	New Jersey, USA	3,792	("SAYREVILLE: 4,403 kWp – Ciel et Terre International," n.d.)
Otae Floating Solar Farm	3	Sangju Province, South Korea	19,432	("Sangju, South Korea I SMA Solar," n.d.)
Godley Reservoir Floating farm	3	Greater Manchester, UK	3,250	("How a floating solar farm powers water treatment – New Civil Engineer," n.d.)
Kato-shi Solar Farm	2.8	Takaoka, Kansai, Japan	11,256	("KATO SHI: 2,878 kWp – Ciel et Terre International," n.d.)
CMIC Pond Floating Solar Farm	2.8	Cambodia	7,768	("CMIC POND: 2,835 kWp – Ciel et Terre International," n.d.)
Hyoshiga IKE Floating Solar Farm	2.7	Hyogo, Japan	10,010	("HYOSHIGA IKE: 2,703 kWp – Ciel et Terre International," n.d.)
Iwano IKE Floating Solar Farm	2.6	Okayama, Japan	8,800	("IWANO IKE: 2,596 kWp – Ciel et Terre International," n.d.)

(*Continued*)

TABLE 8.2 (*Continued*)
Megawatt Power Floating Solar Farms Around the World

Plant Name	Capacity in MW	Location	No of Installed Panels	References
Tsuga IKE Solar Farm	2.45	Mie, Japan	9,072	("TSUGA IKE: 2,449 kWp – Ciel et Terre International," n.d.)
Higashi OTA Floating Solar Farm	2.44	Kagawa, Japan	9,020	("HIGASHI OTA: 2,435 KWP – Ciel et Terre International," n.d.)
Togawa IKE Floating Solar Farm	2.36	Hyogo. Japan	8,733	("TOGAWA IKE: 2,359 KWP – Ciel et Terre International," n.d.)
Sakasama IKE Solar Farm	2.3	Kasai, Japan	9,072	("SAKASAMA IKE: 2,313 kWp – Ciel et Terre International," n.d.)
Hanaoka IKE Floating Solar Farm	2.29	Hyogo, Japan	6,107	("HANAOKA IKE: 2,290 KWP – Ciel et Terre International," n.d.)
Azalaelaan Floating Solar Farm	1.85	Netherland, Europe	6,150	("Azalealaan: 1,845 kWp – Ciel et Terre International," n.d.)
Hikuni IKE Floating Solar Farm	1.30	Hyogo, Japan	3,492	("HIKUNI IKE: 1,308 kWp – Ciel et Terre International," n.d.)
Sobradinho Floating Solar Farm	1	Bahia, Brazil	3,792	("SOBRADINHO: 1,005 kWp – Ciel et Terre International," n.d.)

8.6   FLOATING RENEWABLE-POWERED BUOYS

Floating buoys are employed by both civil and military organizations for a variety of operations ranging from meteorological and pollution monitoring to surveillance at harbor and sea. These buoys have evolved over a period of time: from battery-powered buoys to automated hybrid-powered buoys utilizing a fuel cell source coupled with renewable energy source, namely, solar and tidal energy. Most of the times, these buoys are placed where the surroundings are inhabited, and the data from these buoys (if placed at a distant water body) is often collected on a periodic basis. With the advent of IOT and advanced integrated sensor network, the captured data is readily available on a real-time basis for further analysis (Figure 8.6).

The US Navy in conjunction with Aegeus Technologies has deployed a series of intelligent renewable-powered buoys with integrated high-tech sensors and wireless communication devices. These buoys are primarily deployed for continuous video

FIGURE 8.5   Clockwise from bottom (a) Floating solar farm at Banasurasagar Dam, Kerala (Photo credit: Cdr R Raajiv). (b) Middle Left-85MW Azalaelaan floating Solar farm, the Netherlands, Europe ("Azalealaan: 1,845 kWp – Ciel et Terre International," n.d.). (c) Middle Right-17 MW Piolenc Floating solar farm, France ("O'MEGA1: first floating solar power plant in France | Bouygues Energies & Services," n.d.) (d) Top – 70MW Anhui CECEP floating ("ANHUI CECEP: 70,005 kWp – Ciel et Terre International," n.d.).

surveillance, intermediary modems, sensing oil spills, collecting environmental data, and to act as radiation hazard sensors ("Buoy System Tracks Ships in Real Time – Aerospace & Defense Technology," n.d.). These sensors integrated with IoT gateway devices will enable real-time monitoring of data. In addition to the standard set of sensors, these buoys have in-built GPS and position sensors to accurately monitor position and drift in the area and can be custom reconfigured as per the deployment area. Commercial solar-powered unmanned vessels are being developed by

FIGURE 8.6 Left to right (a) Renewable-powered military buoy for Maritime Surveillance ("Buoy System Tracks Ships in Real Time – Aerospace & Defense Technology," n.d.). (b) Commercial autonomous self-powered boat ("Autonomous self-powered boats 'could create IoT of the ocean' – Energy Live News," n.d.) (photo credit: Open ocean Robotics).

open ocean robotics, which will house an array of sensors to monitor ocean environment ("Autonomous self-powered boats 'could create IoT of the ocean' – Energy Live News," n.d.). The completely renewable-powered vessel is scheduled to undertake a 5000-km journey from Canada to Ireland.

8.7   CASE STUDY: CATALINA SEA RANCH (CSR)

The Catalina Sea Ranch (CSR) is located along the Californian coast of the USA. The Ranch is the first legitimate offshore aqua-cultivation facility developed for the growth of mussels ("How to Build a Data-Capturing Internet of Things Platform for the Ocean – AgFunderNews," 2016). The growth of the mussels depends on the nutrients available from the sea shelf by means of phytoplankton available in the region. Because of the absorption due to marine pollution, the oceanic region tends to get acidic, which in turn will threaten the sustainability of marine phytoplankton, and thus in turn the mussel growth and culture. Monitoring of sea water parameters has become extremely critical on a real-time basis. A trademark IOT technology known as ocean IOT has been implemented on a buoy towards real-time monitoring of the area. The floating buoy is a hybrid-powered one using solar and wind energies to generate power and transmit data on a real-time basis. The NOMAD buoy is provisioned by Integrated Ocean Observatory System (IOOS) of the National Oceanic and Atmospheric Administration (NOAA).

The anchored NOMAD buoy is equipped with various equipment sensors and positioning systems (to monitor in horizontal and vertical axes). The body of the boat buoy is made of aluminum and is 10 feet wide and 20 feet long with equipment arranged below the battery deck. NOMAD is powered by 04×145 W M/s Kyocera solar panels and a Vertical Axis Wind turbine (VAWT) ("Kyocera Solar Powers NOMAD Sea Buoy – Solar Novus Today," 2016). The renewable-powered wireless sensors transmit data in real time and are analyzed on a day-to-day basis by the experts from the cloud in a remote location. IOT sensors and real-time data analysis will aid experts in allocating geographical locations in the sea for other activities to

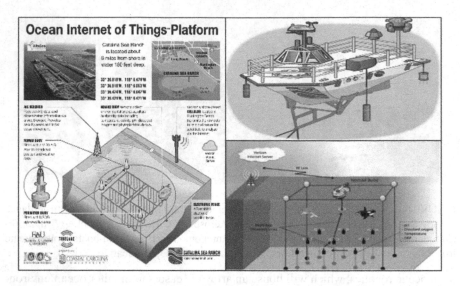

FIGURE 8.7   Clockwise from bottom right (Cruver, 2016) (a) NOMAD buoy communicating with moored sensors measuring pH and dissoleved oxygen temperature. (b) Amalgamation of other buoys with renewable-powered NOMAD buoy and its shore-based receiver for cloud computing. (c) Pictorial representation of NOMAD buoy.

avoid coexisting with the existing one. With amalgamation of sensors ("Kyocera Solar Powers Innovative Marine Big Data and Ocean Internet of Things Platform – The Leading Solar Magazine In India," n.d.) to monitor pH, dissolved oxygen, salinity, sea water temperature, humidity, pressure, wind speed, weather, and an electronic perimeter infused with Automatic Identification System (AIS), the shore facility (located at 10 miles) can not only remotely monitor the data but can also plan and implement corrective measures and enforce rules based on the analyzed data (Figure 8.7).

8.8   RENEWABLE-POWERED MARINE VESSELS

Towards unified management of remote assets on ocean surface, the global satellite giant company Iridium has come up with a new solar-powered satellite IoT device. This will enable customers to remotely monitor their assets even from their mobile devices. In order to combat global emission standards as well as to switch over to renewable power, Eco Marine Power has revealed a project Aquarius concept ship in Japan ("Eco Marine Power's wind and solar-powered ship unveiled in Japan," n.d.). The concept ship utilizes wind and solar powers to energize the ship at sea as well as alongside the harbor. With implementation of IoT-enabled devices real-time monitoring of the vessel is feasible, which can enable the team to gather efficient data. The data collected from onboard can be utilized to efficiently use power based on the ambient conditions in the ocean and thus increase the overall efficiency of energy consumption (Figure 8.8).

Shipping containers form the backbone of the global supply chain business. The efficiency of the whole process lies between the time of supply and delivery of finished

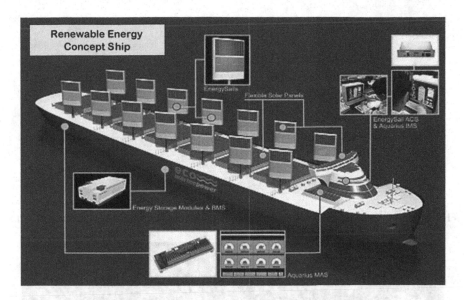

FIGURE 8.8   Eco marine power concept ship ("Aquarius Eco Ship | Eco Marine Power," n.d.).

goods between the manufacturers and the consumers. IoT-enabled shipping will enable a sea change towards identification and tracking of container on a real-time basis. On average, a container ship will carry anywhere between 15,000 and 23,000 TEU (twenty-foot equivalent unit) containers onboard. With the introduction of IoT-enabled smart container, it is now possible to micromanage data pertinent to a single container on a real-time basis. IoT-enabled containers can transmit the following data:

   a. Location Sensors – to identify the precise location of container
   b. Humidity Sensors – to monitor the condition of perishable goods inside
   c. Door-Lock Sensors – to have accountability of door open and shut
   d. Motion Sensors – to prevent theft and unwanted movement en-route

8.9   CASE STUDY: SMART CONTAINERS

A Rotterdam-based company (M/s We Are 42) has come up with the idea of a smart container. An IoT-enabled solar-powered container housing a variety of sensors housed on a container was studied over a 2-year period for assessment since 2018. It has been established that the proof of concept was highly successful and can be introduced into shipping for betterment (Figure 8.9).

The Container 42 featured sensors that could record the temperature inside and outside the container, humidity, air pollution level, vibration, and motion; in addition, it could transmit time-lapse images to the network ("Technology – We Are 42," n.d.). The lock sensors provide information about the time and other additional data about the time, person, and location of the container unlock. The roof of the container is fitted with Solar panels that will generate power for the IoT sensors and their functionalities. A sample screenshot of container 42 monitoring is shown in Figure 8.10.

FIGURE 8.9   Solar-powered IoT enabled smart container ("Technology – We Are 42," n.d.).

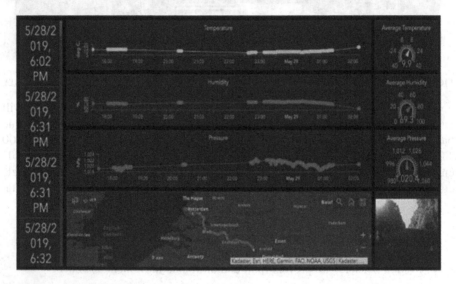

FIGURE 8.10   Screenshot of monitoring of solar-powered IoT enabled Smart Container ("IoT-enabled shipping containers sail the high seas improving global supply chains. | Network World," n.d.)

8.10   EFFICIENT SOLAR-POWERED MARINE VESSELS

Marine vessels (as shown in Figure 8.11) powered completely by renewable source as well as hybridized power combining conventional fuel and renewable energy has been in use for considerable time, and with the onset of big data and predictive analysis induced by IoT platforms these vessels will stand to gain much more efficient methods of operation. IoT will enable not only real-time monitoring but also will contribute to efficient route planning, power consumption, and reduction in system downtime by inducing timely automated maintenance routines during the course of operation of these vessels.

FIGURE 8.11 Clockwise from bottom (a) Solar panels installed onboard Auriga Leader Cargo carrier ("Toyota's Auriga Leader transport ship gets hybridized [w/video] | Autoblog," n.d.). (b) Solar-powered ferry Aditya in Vaikom, Kerala ("'ADITYA' – India's first solar powered boat – Photo Gallery – State Water Transport Department, Government of Kerala, India," n.d.). (c) Solar-powered yacht – Turanor Planet Solar ("World's largest all-solar-powered boat shines in NYC | News | Eco-Business | Asia Pacific," n.d.).

8.11 CONCLUSION

Solar energy has made tremendous progress and has achieved significant technological maturity over the years. The sanctions imposed by IMO in the maritime domain have caused a paradigm shift in the thought process of marine propulsion, and the focus has shifted toward implementation of green energy and reduction of noxious emissions. Floating solar farms are considered as a stupendous technological advancement wherein the benefits outweigh the constraints. These floating solar farms combined with IoT will definitely make a tremendous impact during the forthcoming years. The technology has shown tremendous growth potential in the commercial and military sectors. The IoT sector is expanding at a steady pace and has been alternatively known in the marine sector as IoT for the Sea, Ocean of Things (OoT), and Internet of Underwater Things (IoUT). Observational devices deployed in the oceans around the world include position floats, moored buoys, weather and wave monitoring buoys, and navigation buoys; these devices are much more efficient than a routine survey vessel undertaking measurements in an area at a given fixed time. IoT combined with satellite ability will enable greater mapping of ocean area. The sustainability of a deployed device can also be micromanaged based on the energy planning undertaken from the historical data send by these devices. The low cost of these IoT devices combined with renewable powering makes it an attractive option for future deployments.

REFERENCES

About IRENA. (n.d.). Retrieved June 27, 2020, from https://www.irena.org/aboutirena

"ADITYA" – India's first solar powered boat – Photo Gallery – State Water Transport Department, Government of Kerala, India. (n.d.). Retrieved September 6, 2020, from https://www.swtd.kerala.gov.in/pages-en-IN/pg-aditya.php

AGONGDIAN: 9,994 kWp – Ciel et Terre International. (n.d.). Retrieved September 4, 2020, from https://www.ciel-et-terre.net/project/agongdian-9994-kwp/

Andoni, M., Robu, V., Flynn, D., Abram, S., Geach, D., Jenkins, D., ... Peacock, A. (2019). Blockchain technology in the energy sector: A systematic review of challenges and opportunities. *Renewable and Sustainable Energy Reviews, 100*(February 2018), 143–174. doi:10.1016/j.rser.2018.10.014

ANHUI CECEP: 70,005 kWp – Ciel et Terre International. (n.d.). Retrieved September 4, 2020, from https://www.ciel-et-terre.net/project/anhui-cecep-70005-kwp/

ANHUI GCL: 32,686 kWp – Ciel et Terre International. (n.d.). Retrieved September 4, 2020, from https://www.ciel-et-terre.net/project/anhui-gcl-32686-kwp/

Aquarius Eco Ship | Eco Marine Power. (n.d.). Retrieved October 14, 2020, from https://www.ecomarinepower.com/en/aquarius-eco-ship

Asia and Pacific. (2020). Retrieved June 27, 2020, from https://www.irena.org/asiapacific

Autonomous self-powered boats "could create IoT of the ocean" – Energy Live News. (n.d.). Retrieved September 5, 2020, from https://www.energylivenews.com/2019/06/17/autonomous-self-powered-boats-could-create-iot-of-the-ocean/

Azalealaan: 1,845 kWp – Ciel et Terre International. (n.d.). Retrieved September 5, 2020, from https://www.ciel-et-terre.net/project/azalealaan-1845-kwp/

Baker, T., Asim, M., MacDermott, Á., Iqbal, F., Kamoun, F., Shah, B., ... Hammoudeh, M. (2020). A secure fog-based platform for SCADA-based IoT critical infrastructure. *Software – Practice and Experience, 50*(5), 503–518. doi:10.1002/spe.2688

BayWa r.e. builds Netherland's biggest Floating Solar Farm – BayWa r.e. (n.d.). Retrieved September 5, 2020, from https://www.baywa-re.com/en/news/details/baywa-re-builds-netherlands-biggest-floating-solar-farm/

Bensaleh, M. S., Saida, R., Kacem, Y. H., & Abid, M. (2020). Wireless sensor network design methodologies: A survey. *Journal of Sensors, 2020*. doi:10.1155/2020/9592836

BP. (2020). Statistical Review of World Energy, 2020 | 69th Edition. *Bp*, 66. Retrieved from https://www.bp.com/content/dam/bp/business-sites/en/global/corporate/pdfs/energy-economics/statistical-review/bp-stats-review-2020-full-report.pdf

Buoy System Tracks Ships in Real Time – Aerospace & Defense Technology. (n.d.). Retrieved September 5, 2020, from https://www.aerodefensetech.com/component/content/article/adt/features/application-briefs/15724

Can Floating Solar Plants Break Free From the Trappings of High Maintenance? (n.d.). Retrieved September 2, 2020, from https://www.moxa.com/en/articles/can-floating-solar-plants-break-free-from-the-trappings-of-high-maintenance

Carlos-Mancilla, M., López-Mellado, E., & Siller, M. (2016). Wireless sensor networks formation: Approaches and techniques. *Journal of Sensors, 2016*. doi:10.1155/2016/2081902

Cheruvu, S., Kumar, A., Smith, N., & Wheeler, D. M. (2020). *Demystifying Internet of Things Security. Demystifying Internet of Things Security*. doi:10.1007/978-1-4842-2896-8

Chlorofluorocarbons and Ozone Depletion – American Chemical Society. (n.d.). Retrieved October 17, 2020, from https://www.acs.org/content/acs/en/education/whatischemistry/landmarks/cfcs-ozone.html

Choi, Y. K. (2014). A study on power generation analysis of floating PV system considering environmental impact. *International Journal of Software Engineering and Its Applications, 8*(1), 75–84. doi:10.14257/ijseia.2014.8.1.07

CMIC POND: 2,835 kWp – Ciel et Terre International. (n.d.). Retrieved September 4, 2020, from https://www.ciel-et-terre.net/project/cmic-pond-2835-kwp/

Colcomb, K., Rymell, M., & Lewis, A. (2006). Very heavy fuel oils: risk analysis of their transport in UK waters. *VLIZ Special Publication*, 40–51. Retrieved from http://citeseerx.ist.psu.edu/viewdoc/download?doi=10.1.1.126.8487&rep=rep1&type=pdf#page=46%5Cnhttp://www.vliz.be/imisdocs/publications/103728.pdf

Comer, B. (2019). Transitioning away from heavy fuel oil in Arctic shipping, 12pp.

Cruver, P. (2016). How to Build a Data-Capturing Internet of Things Platform for the Ocean – AgFunderNews. Retrieved October 14, 2020, from https://agfundernews.com/how-to-build-a-data-capturing-internet-of-things-for-the-ocean5478.html

Demirbas, A. (2007). Importance of biodiesel as transportation fuel. *Energy Policy, 35*(9), 4661–4670. doi:10.1016/j.enpol.2007.04.003

Eco Marine Power's wind and solar-powered ship unveiled in Japan. (n.d.). Retrieved September 5, 2020, from https://www.ship-technology.com/news/eco-marine-power-ship/

Edeland, C., & Mörk, T. (2018). Blockchain Technology in the Energy Transition an Exploratory Study on How Electric Utilities Blockchain Technology in the Energy Transition. *Trita-Itm-Ex Nv -2018:78, Independen*, 141. Retrieved from http://kth.diva-portal.org/smash/get/diva2:1235832/FULLTEXT01.pdf%0Ahttp://urn.kb.se/resolve?urn=urn:nbn:se:kth:diva-232668

Fersi, G. (2015). A distributed and flexible architecture for Internet of Things. *Procedia Computer Science, 73*(Awict), 130–137. doi:10.1016/j.procs.2015.12.058

GCL JINING: 6,776 kWp – Ciel et Terre International. (n.d.). Retrieved September 4, 2020, from https://www.ciel-et-terre.net/project/gcl-jining-6776-kwp/

Han, J., Song, W., Gozho, A., Sung, Y., Ji, S., Song, L., … Zhang, Q. (2020). LoRa-Based smart IoT application for smart city: An Example of Human Posture Detection. *Wireless Communications and Mobile Computing, 2020*. doi:10.1155/2020/8822555

HANAOKA IKE: 2,290 KWP – Ciel et Terre International. (n.d.). Retrieved September 4, 2020, from https://www.ciel-et-terre.net/project/hanaoka-ike-2290-kwp/

HIGASHI OTA: 2,435 KWP – Ciel et Terre International. (n.d.). Retrieved September 5, 2020, from https://www.ciel-et-terre.net/project/higashi-ota-2435-kwp/

HIKUNI IKE: 1,308 kWp – Ciel et Terre International. (n.d.). Retrieved September 4, 2020, from https://www.ciel-et-terre.net/project/hikuni-ike-1308-kwp/

How a Floating Solar Farm Powers Water Treatment – New Civil Engineer. (n.d.). Retrieved September 4, 2020, from https://www.newcivilengineer.com/innovative-thinking/how-a-floating-solar-farm-powers-water-treatment-16-04-2019/

How to Build a Data-Capturing Internet of Things Platform for the Ocean – AgFunderNews. (2016). Retrieved July 31, 2020, from https://agfundernews.com/how-to-build-a-data-capturing-internet-of-things-for-the-ocean5478.html

Huainan: Largest Floating Solar Farm in the World | Planète Énergies. (n.d.). Retrieved September 4, 2020, from https://www.planete-energies.com/en/medias/close/huainan-largest-floating-solar-farm-world

Hunzinger, R. (2017). Scada fundamentals and applications in the IoT. *Internet of Things and Data Analytics Handbook*, 283–293. doi:10.1002/9781119173601.ch17

HYOSHIGA IKE: 2,703 kWp – Ciel et Terre International. (n.d.). Retrieved September 4, 2020, from https://www.ciel-et-terre.net/project/hyoshiga-ike-2703-kwp/

International Maritime Organization. (n.d.). Retrieved September 5, 2020, from http://www.imo.org/en/Pages/Default.aspx

International Renewable Energy Agency – IRENA. (2019). Innovation landscape brief: Behind-the-meter batteries. *Irena 2019*, 24.

IoT-Enabled Shipping Containers Sail the High Seas Improving Global Supply Chains. | Network World. (n.d.). Retrieved September 5, 2020, from https://www.networkworld.com/article/3432170/iot-enabled-shipping-containers-sail-the-high-seas-improving-global-supply-chains.html

IWANO IKE: 2,596 kWp – Ciel et Terre International. (n.d.). Retrieved September 5, 2020, from https://www.ciel-et-terre.net/project/iwano-ike-2596-kwp/

KATO SHI: 2,878 kWp – Ciel et Terre International. (n.d.). Retrieved September 4, 2020, from https://www.ciel-et-terre.net/project/kato-shi-2878-kwp/

Kim, S. H., Yoon, S. J., & Choi, W. (2017). Design and construction of 1MW class floating PV generation structural system using FRP members. *Energies*, *10*(8), 1–14. doi:10.3390/en10081142

Kim, S. M., Oh, M., & Park, H. D. (2019). Analysis and prioritization of the floating photovoltaic system potential for reservoirs in Korea. *Applied Sciences (Switzerland)*, *9*(3). doi:10.3390/app9030395

Kyocera Solar Powers Innovative Marine Big Data and Ocean Internet of Things Platform – The Leading Solar Magazine in India. (n.d.). Retrieved September 2, 2020, from https://www.eqmagpro.com/kyocera-solar-powers-innovative-marine-big-data-and-ocean-internet-of-things-platform/

Kyocera Solar Powers NOMAD Sea Buoy – Solar Novus Today. (2016). Retrieved July 31, 2020, from https://www.solarnovus.com/kyocera-solar-powers-nomad-sea-buoy_N10331.html

Lee, C. K. M., Cao, Y., & Ng, K. K. H. (2017). Big data analytics for predictive maintenance strategies. *Supply Chain Management in the Big Data Era*, (June), 50–74. doi:10.4018/978-1-5225-0956-1.ch004

Menon, R. R. (2014). Exploration and production issues in South Asia. *Journal of Unconventional Oil and Gas Resources, 6*. doi:10.1016/j.juogr.2013.09.003

Nebey, A. H., Taye, B. Z., & Workineh, T. G. (2020). GIS-based irrigation dams potential assessment of floating solar PV system. *Journal of Energy, 2020*, 1–10. doi:10.1155/2020/1268493

O'MEGA1: First Floating Solar Power Plant in France | Bouygues Energies & Services. (n.d.). Retrieved September 5, 2020, from https://www.bouygues-es.com/energy/omega1-first-floating-solar-power-plant-france

Oh, S.-C., Kim, M.-S., Park, Y., Roh, G.-T., & Lee, C.-W. (2017). Implementation of blockchain-based energy trading system. *Asia Pacific Journal of Innovation and Entrepreneurship*, *11*(3), 322–334. doi:10.1108/apjie-12-2017-037

PEI COUNTY: 9,982 kWp – Ciel et Terre International. (n.d.). Retrieved September 4, 2020, from https://www.ciel-et-terre.net/project/pei-county-9982-kwp/

Project Briefing: New Floating Solar Concepts at BayWa r.e.'s Sekdoorn | PV Tech. (n.d.). Retrieved September 5, 2020, from https://www.pv-tech.org/guest-blog/project-briefing-new-floating-solar-concepts-at-baywa-r.e.s-sekdoorn

Ray, P. P. (2018). A survey on Internet of Things architectures. *Journal of King Saud University – Computer and Information Sciences, 30*(3), 291–319. doi:10.1016/j.jksuci.2016.10.003

Renewable Energy Agency International IRENA. (2019). *Renewable Energy Market Analysis: Latin America. Irena.*

Sajid, A., Abbas, H., & Saleem, K. (2016). Cloud-assisted IoT-based SCADA systems security: A review of the state of the art and future challenges. *IEEE Access, 4*(c), 1375–1384. doi:10.1109/ACCESS.2016.2549047

SAKASAMA IKE: 2,313 kWp – Ciel et Terre International. (n.d.). Retrieved September 4, 2020, from https://www.ciel-et-terre.net/project/sakasama-ike-2313-kwp/

Sangju, South Korea | SMA Solar. (n.d.). Retrieved September 4, 2020, from https://www.sma.de/en/products/references/sangju-south-korea.html

Saritha, S., & Sarasvathi, V. (2019). A study on application layer protocols used in IoT. *2017 International Conference on Circuits, Controls, and Communications (CCUBE)*, 155–159. doi:10.1109/ccube.2017.8394143

SAYREVILLE: 4,403 kWp – Ciel et Terre International. (n.d.). Retrieved September 4, 2020, from https://www.ciel-et-terre.net/project/sayreville-4403-kwp/

Sethi, P., & Sarangi, S. R. (2017). Internet of Things: Architectures, protocols, and applications. *Journal of Electrical and Computer Engineering, 2017.* doi:10.1155/2017/9324035

Shahzad, Aa., Kim, Y. G., & Elgamoudi, A. (2017). Secure IoT platform for industrial control systems. *2017 International Conference on Platform Technology and Service, PlatCon 2017- Proceedings*, 0–5. doi:10.1109/PlatCon.2017.7883726

SOBRADINHO: 1,005 kWp – Ciel et Terre International. (n.d.). Retrieved September 4, 2020, from https://www.ciel-et-terre.net/project/sobradinho-1005-kwp/

Stead, C., Wadud, Z., Nash, C., & Li, H. (2019). Introduction of biodiesel to rail transport: Lessons from the road sector. *Sustainability (Switzerland), 11*(3), 1–20. doi:10.3390/su11030904

SUGU 1: 4,023 kWp – Ciel et Terre International. (n.d.). Retrieved September 4, 2020, from https://www.ciel-et-terre.net/project/sugu-1-4023-kwp/

Technology – We Are 42. (n.d.). Retrieved September 5, 2020, from https://weare42.io/technology/

Thangavel, S. (2018). Ozone Layer Depletion and Its Effects : A Review, (May). https://doi.org/10.7763/IJESD.2011.V2.93

TOGAWA IKE: 2,359 KWP – Ciel et Terre International. (n.d.). Retrieved September 5, 2020, from https://www.ciel-et-terre.net/project/togawa-ike-2359-kwp/

Toyota's Auriga Leader Transport Ship Gets Hybridized [w/video] | Autoblog. (n.d.). Retrieved September 6, 2020, from https://www.autoblog.com/2011/06/26/toyotas-auriga-leader-transport-ship-gets-hybridized-w-video/

TSUGA IKE: 2,449 kWp – Ciel et Terre International. (n.d.). Retrieved September 4, 2020, from https://www.ciel-et-terre.net/project/tsuga-ike-2449-kwp/

UMENOKI: 7,550 kWp – Ciel et Terre International. (n.d.). Retrieved September 4, 2020, from https://www.ciel-et-terre.net/project/umenoki-7550-kwp/

UNEP's India Solar Loan Programme Wins Energy Globe. (2007). Retrieved June 27, 2020, from https://www.solutions-site.org/node/258

UNEP – UN Environment Programme. (n.d.). Retrieved September 5, 2020, from https://www.unenvironment.org/

What Is SCADA? Supervisory Control and Data Acquisition. (n.d.). Retrieved September 2, 2020, from https://inductiveautomation.com/resources/article/what-is-scada

World's Biggest Floating Solar Farm Powers Up Outside London | Environment | The Guardian. (n.d.). Retrieved September 4, 2020, from https://www.theguardian.com/environment/2016/feb/29/worlds-biggest-floating-solar-farm-power-up-outside-london

World's Largest All-Solar-Powered Boat Shines in NYC | News | Eco-Business | Asia Pacific. (n.d.). Retrieved September 6, 2020, from https://www.eco-business.com/news/worlds-largest-all-solar-powered-boat-shines-nyc/

Xu, G., Shi, Y., Sun, X., & Shen, W. (2019). Internet of Things in marine environment monitoring: A review. *Sensors (Switzerland)*, *19*(7), 1–21. doi:10.3390/s19071711

Yamakura Dam Welcomed Floating Solar Panels on Its Surface – Ciel et Terre International. (n.d.). Retrieved September 4, 2020, from https://www.ciel-et-terre.net/yamakura-dam-welcomed-floating-solar-panels-on-its-surface/

Yassein, M. B., Shatnawi, M. Q., & Al-Zoubi, D. (2016). Application layer protocols for the Internet of Things: A survey. *Proceedings -2016 International Conference on Engineering and MIS, ICEMIS 2016*. doi:10.1109/ICEMIS.2016.7745303

Yusoff, M. N. A. M., Zulkifli, N. W. M., Sukiman, N. L., Chyuan, O. H., Hassan, M. H., Hasnul, M. H., ... Zakaria, M. Z. (2020). Sustainability of palm biodiesel in transportation: A review on biofuel standard, policy and international collaboration between Malaysia and Colombia. *Bioenergy Research*. doi:10.1007/s12155-020-10165-0

Zailani, S., Iranmanesh, M., Hyun, S. S., & Ali, M. H. (2019). Barriers of biodiesel adoption by transportation companies: A case of Malaysian transportation industry. *Sustainability (Switzerland)*, *11*(3). doi:10.3390/su11030931

Zerefos, C., Contopoulos, G., & Gregory, S. (Eds.). (2009). Twenty years of ozone decline. In *Proceedings of the Symposium for the 20th Anniversary of the Montreal Protocol* (p. 470). Springer. doi:10.1007/978-90-481-2469-5

Zourmand, A., Kun Hing, A. L., Wai Hung, C., & Abdulrehman, M. (2019). Internet of Things (IoT) using LoRa technology. *2019 IEEE International Conference on Automatic Control and Intelligent Systems, I2CACIS 2019- Proceedings*, (June), 324–330. doi:10.1109/I2CACIS.2019.8825008

9 Deep Learning Approach towards Solar Energy Forecast

Anitya Kumar Gupta
Illinois Institute of Technology, Chicago, Illinois

Vikas Pandey
Babu Banarasi Das University, Lucknow, India

Akhilesh Sharma
NERIST, Nirjuli, India

Safia A. Kazmi
ZHCET, Aligarh Muslim University, Aligarh, India

CONTENTS

9.1   INTRODUCTION

The world is moving towards sustainable renewable power sources (RES). This has driven the turn of events to the need for photovoltaic (PV) boards. The cost of delivering power from PV boards has reduced over the years, while expanding the vitality transformation proficiency. According to the statisticians, it is seen that an improvement of 5% occurred during the year 2017. This improvement may be due to a number of factors like proper harnessing of energy during the season, efficient design, etc.

Nevertheless, since PV board vitality yield depends upon weather conditions, such as overcast spread and sun-powered irradiance, duration of the solar radiation, and the angle of incidence of sun's radiations, the vitality yield of the PV boards is unreliable. To understand and deal with the yield fluctuation is of interest for a few enthusiasts in the vitality field. Temporarily (up to 5 hours), a transmission framework administrator is keen on the vitality yield from PV boards to locate sufficient parity for the entire matrix, since over- and under-creating power frequently brings additional penalty charges [1]. On the other side of the range, power dealers are keen on long time skylines, normally, one day-ahead conjectures, since most power is exchanged on the day-ahead market. Thus, the benefit of these tasks depends on the capacity to figure out fluctuations in sun-powered PV boards vitality yield precisely.

Corresponding to the expanded interest in PV power anticipating arrangements, the methods for anticipating with the assistance of AI (ML) have, over time, become popular compared with conventional time arrangement-prescient models. Although ML procedures are the same old thing, the enhanced computational limit and the higher accessibility of value information have made the methods helpful for anticipating [2] and it can be considered to be one of the vital procedures in the specific arena in the future.

The fundamental goal is to benchmark diverse estimating strategies of sun-based PV board vitality yields. Towards this end, AI and time arrangement procedures can be utilized to powerfully become familiar with the connection between various climatic conditions and the vitality yield of PV frameworks. There are four ML procedures that are the benchmark to customary time arrangement techniques on PV framework information in existing establishments. This, likewise, required an examination of important building procedures, which can be utilized to expand the general prediction accuracy [3] of the PV system.

TABLE 9.1

Existing Solar Forecast Accuracy

Authors	Error Rate (ER) (%)
Energy & Metro System	19
Yang (2012)	36
Remund (2008)	20
Zhang (2013)	15

For the PV system, it is really challenging to handle the system so as to improve the solar energy issue with least error rate. With the world in the AI revolution now, all the scientists have the potential to work and bring out the best solution, as can be seen in Table 9.1 [4], which provides the insights of the different authors and gives their research results with the significant value of error.

Sun-oriented boards, wind turbines, and maritime force are one of the groups that falls into this class [4]. As the non-dispatchable vitality sources are expanding, the force created by these inexhaustibles should be considered in the activities and arranging of Autonomous Force Makers/Administrators, and Adjusting Specialists. Since the dependable execution of the mass framework relies upon the adjusting of a persistently differing load with equivalent measures of age, information on the heap and inexhaustible age in front of time is basic for the financially ideal and actually attainable dispatch of age sources. Framework administrators need to guarantee that they have adequate assets to oblige huge up or down inclines because of the stochastic nature of sustainable age to maintain the framework balance. Systems for working the force framework depend on precise figures of conditions to come at various figure time skylines, i.e., seconds, minutes, hours, or/and days ahead [5]. Along these lines, progressively exact figures of sustainable power sources will decrease the lopsidedness charges and penalties and give a serious edge to the progressively and day-ahead vitality advertise exchanging.

Conventional sun-oriented estimate strategies use point-wise climate conjecture data at the area of the sun-oriented homestead. Such point-wise data function admirably on bright days, or cloudy conditions, where the irradiance coming from the sky is generally even. Be that as it may, such a strategy endures instances of incompletely overcast conditions where the overcast spreads portions of the sky what is more, with steady developments [6]. Since it is exceptionally hard to accurately predict the advancement and dispersal of clouds, along these lines the conjecture of the episode irradiance on the specific area of the sunlight-based board is not exact.

The main focus is on a couple of colors, red and blue, as indicated in Figure 9.1. Blue color provides the overview of the low solar irradiance and red color is the high level of solar irradiance. Point-wise climate data show that it is bright conditions at the ranch. Be that as it may, the genuine sun-powered vitality creation is not as high, true to form. At the point when one plots out the guide of the sun-oriented irradiance esteem around the homestead for the term of the entire hour, it is found there are mists (depicted by pale blue shading) in the region of this sunlight-drenched ranch and headed for the sun-based ranch. Given the uncertainty in

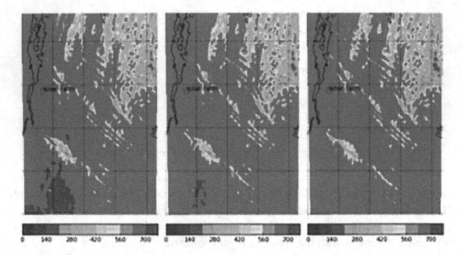

FIGURE 9.1   Forecasting issue during the cloudy weather.

climate estimating, the real appearance time of cloud, as given in the guide, may not be precise either. Thus, the results may not be accurate. Consequently, for the undertaking of sun-based estimating, it is important to demonstrate the altitudinal and real time example of irradiance to figure out the movement of cloud that influences the sunlight-based vitality creation [4].

9.2   LITERATURE REVIEW

This section gives a concise overview of the innovation space, in particular sun-powered force and sun-oriented force vitality stockpiling. Research applicable to the innovations is assessed to create essential cosmology diagrams of the area to build the inquiry strings used to question online patent and writing databases. An iterative methodology is utilized to look through the references and earlier expressions to enhance the quest for extra licenses and exploration papers.

9.2.1   STORAGE ENERGY AND POWER CELLS

As innovation develops, the item life span moves in the developmental stage. This necessitates a quick, expanding interest for hardware and administrations. Sustainable power sources are more of a costly option with continuous worries about the productivity, cost, and usage across far-reaching electrical lattice foundations. Most sustainable power source innovations are in the late starting phase of the item life span but as time progresses, scientists are finding it difficult and losing their keen interest. This sort of market reaction is frequently called the "Bombard Impacting" since noteworthy capital ventures have been made in non-sustainable power source offices that are not completely deteriorated and can work for some more decades [5]. Sustainable power source, for example, wind and sunlight-based force cannot deliver power dependably with flow innovation since power creation rates change with seasons, months, and even on a weekly basis or on specific days, or even within a day.

The commercial center calls for huge scope and moderate answers for lighten fluctuating which tends to yield and give strategies to store abundance creation for later utilization [7]. Sun-oriented vitality is one of the most widely recognized and well-known wellsprings of clean vitality, and the necessity to understand daylight is an exceptionally straightforward prerequisite contrasted with different arrangements. Direct sunlight-based radiation may have the best potential for enormous scope, which if of use once practical vitality stockpiling innovation is created. [2] audited this, and his thought process gives the necessary information about both the benefits and constraints of sun-based vitality advances.

Concentrating sun-based force (CSP) plants create sunlight-based warm power without ozone-depleting substance discharge and has a vitality innovation with no undesirable effect on environmental alteration. A thermoelectric sunlight-based plant utilizes many components organized in the accompanying request [8]. The primary part of succession is a mirror that is intended to gather sunlight-based radiation and aggregate it at a point of convergence. The subsequent parts in CSP are the collector and the warmth exchanger, which circles heat by the means of the transmission approach after which it will be pushed towards to liquid, (for example, liquid salt or engineered oil) so as to retain the concentrated warmth. The last element comprises of a second warmth exchanger that moves the aggregated warm vitality to an additional liquid (normally steam). The heated liquid coming out of the exchanger is able to move the blades of the turbine, thus generating electrical energy [9].

To diminish the expense in each region needed by photovoltaic (PV) cells, sunlight-based concentrators depend on many numbers of mirrors or mechanical structures that keep moving to allow the light to get diverted towards the concentrator, which helps in tracking the motion of the sun [10]. Sun-based concentrators have burdens as they have to follow the position of the sun and might be influenced by excess heating from the convergence of light and warmth on the sun-powered cells [11]. The upsides of utilizing volume holographic optical components [12] are engaging for inconsequential weight. They are applied in modest sunlight-based concentrator applications. They are relied upon to turn into a significant headway when incorporated into sun-powered boards. [11] introduced a survey of holographic-based sunlight-based concentrators utilizing various constituents.

The physical standards and fundamental favorable circumstances and drawbacks, for example, cool light focus, specific frequency fixation, and the likelihood to actualize latent sun-based following are discussed. Various setups and application techniques are also mentioned in this investigation.

Unlike all sun-powered PV advances, CSP plants are based on steam, and hence use steam turbines that coordinate customary electrical-creating administrations. These plants can be outfitted with non-renewable energy source frameworks to convey extra vitality or to create power during the night or when the sun is hindered by mists [13]. The reflector mirror available in CSPs are grouped into sunlight-based force towers, Fresnel reflectors, authentic dishes, and illustrative troughs. To store heat, these plants may utilize liquid salt, empowering the age of power for a few hours even without daylight. During off-peak hours, their capacity age could be balanced by power request. The force age may be closed down rapidly. The collected warmth can be put away by the liquid salt [14]. The present most progressive CSP frameworks are towers coordinated with two-tank liquid salt warm vitality stockpiling, conveying warm vitality at 565°C for mixing with customary steam, such as by Rankine power cycles. The force towers follow their heredity to the 10-MWe pilot exhibition of the sun oriented which were determined during the period of 90s. The structure brought down the expense of CSP power generation by roughly half over the earlier age of explanatory trough frameworks. Notwithstanding, the abatement in cost of CSP advancements has not kept up with the reduced expense of PV frameworks [15]. It was discovered that the work of a regular battery had more life-cycle costs (LCC) when compared with a progressed profound cycle battery. This helps in determining, for force gracefully framework, utilization of the profound cycle batteries is more beneficial for such independent framework. The siphoned build up stocks joined with a bank of number of batteries had practically half LCC as a traditional battery, making this consolidated choice more cost-serious than the solitary battery alternative.

The most generally adjusted sun-powered cell is built with silicon wafers. This has a record percentage of 90% worldwide [16]. Because of the deficiency of crude materials, the conventional Si wafer sun-oriented cells are unable to satisfy the need and cost prerequisites for quickly developing worldwide arcade. Slim film conductors have become the innovation focal point of new age sun-based cells as these cells do not need a lot of Si materials. There are numerous kinds of slight film sun powered cells, including Ge films (nebulous germanium a-Si, microcrystalline germanium c-Si, stacked a-Si/c-Si), compound semiconductors (copper indium gallium selenide CIS/CIGS, cadmium telluride CdTe), and color refinement sun-oriented cells (DSSC) [16]. Albeit meager, film sun-based cells have low vitality change proficiency, low large-scale manufacturing yield, and significant expenses. There are numerous focal points, for example, material reserve funds, as they may be created utilizing economical glass or plastic substrates. These materials could be tweaked. Moreover, they also bid more noteworthy adaptability for basic applications.

The PV cell contains a pair of cells that utilizes double sun-oriented cells with various assimilation qualities empowering a more extensive scope of the sun-based range to be changed over to vitality. A straightforward titanium oxide (TiO_x) layer isolates. This interfaces the dual cells. The TiO_x layer fills in as the electron shipping. There exists assortment layer for the primary cell and is the establishment that empowers the creation of the subsequent cell to finish the couple cell engineering [17]. The specialized trouble for battery pair is that the current produced should be essentially the same. Also, the flows created by double layers of the battery are difficult to orchestrate. A high fixation PV innovation has gotten worldwide consideration because of the benefits of effective high-power age, a minimum coefficient of temperature, and the possibility to lessen power age costs. The PV frameworks are, much of the time, intended to work. They are interconnected with the electric utility matrix. An inverter is one of the prime parts of the such framework. The force molding unit (PCU) is another significant component in the framework which changes a direct current power into air conditioning power [18]. The power could be easily predictable with the voltage and force necessities of the lattice. This, consequently, quits providing power when the network satisfies the minimum force needed [19].

The power must be utilized as it is delivered; however, it very well may be put away provided that it is changed over to another vitality structure (for example, compound vitality in batteries) or used to siphon water tough where the hydrostatic force is utilized to control turbines. The restriction of sunlight-based force is that the innovation of changing power into storable vitality has not been developed. To conquer the discontinuity issue of sunlight-based force, a capacity medium or vitality transporter is necessary. There are three advances that are right now utilized as practical vitality stockpiling answers for sun-based force, i.e., savvy batteries, warm vitality stockpiling, and hydrogen power modules. To begin with, keen batteries can store vitality created by sun-based boards, which implies there is no hanging tight for daylight before firing up machines or apparatuses. The vitality created during the day can gracefully control around evening time. Warm vitality stockpiling is generally utilized with warm sun-based force plants which produce high temperatures utilizing mirror clusters instead of photovoltaic boards. The put away warmth (such as liquid salt) disintegrates water (H_2O) into steam to actuate the turbine. Actuation of turbine generates electric current, which is required during the darkness [20]. The power modules act as one of the necessary components of a sunlight-based hydrogen vitality cycle where a framework converts H_2O into its constituents, i.e. hydrogen (H_2) and oxygen (O_2). These constituents are additionally put away by an energy component to create power without daylight. Enormous scope vitality stockpiling arrangements are still in their infancy, yet these advancements will significantly impact the sustainable power source industry

A PV stockpiling framework can be partitioned into off-lattice, on-network and half-breed frameworks. The off-network framework, or independent framework, comprises battery bundles, photovoltaic charge with release regulators, off-matrix inverters, and air conditioning/DC converters [21]. The regulator deals with the charging and releasing of the battery, shields the battery from excess charging and totally releasing. The capacity of the off-network inverter is to change over the DC power into air conditioning control. Thus, power is transferred to a framework or a utility matrix. The structure of an independent framework must consider the limit of the battery to be utilized around evening time, realizing the force load, foreseeing overcast days, and deciding the necessities of sun-oriented cell module load up. The structure is more complex and costlier. The run-of-the-mill application is utilized in elevated earth structures like mountain zones, remote islands, or zones deprived of power matrices.

The on-lattice framework comprises a PV, as shown in Figure 9.2, a PV regulator, battery packs, a battery's executive framework, an inverter, a vitality stockpiling unit, and a dispatch control framework [22]. Sun-based boards transform light vitality into power that charges the pack of battery made up of Li. The direct current power is changed over to air conditioning power through the inverter as indicated in the figure. The regulator constantly switches, thereby altering the working condition of the battery pack as indicated by deviations in daylight power and the heap status. The power is transmitted to air condition converter for surefire use or overabundance while the direct current power is transmitted to storage in a pack of batteries of required capacity. At the point when power age cannot fulfill the heap need, the regulator utilizes stored power in the batteries to guarantee the coherence and strength of the framework. The on-matrix inverter framework contains few inverters that convert the DC power from the pack of batteries into an ac voltage of required magnitude and frequency for

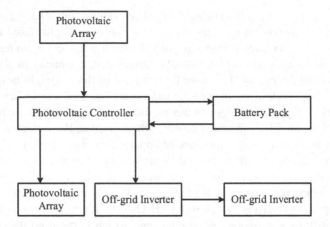

FIGURE 9.2   Key components of the off-grid system.

the client-side low-voltage lattice or for transmitting it to a high-voltage network. The favorable circumstances are protected and basic plan and simple upkeep, with proficient sun-based vitality age that is higher than that of independent frameworks.

An independent PV framework requires vitality stockpiling to gracefully constant vitality when there is lacking or no sun-based emission. The Valve-Controlled Lead Corrosive (VRLA) batteries are now and then utilized yet providing an enormous eruption of current, for example, engine startup debases the battery plates and can annihilate the battery. A technique for providing a lot of consistent current is to join VRLA batteries with a bank of supercapacitors to frame a half and half stockpiling framework where these capacitors deliver moment capacity to the heap [23,24]. Suggested a sort of sun-based battery material, called as 2D cyanimide-functionalized polybetaine imide (NCN-PHI). This material can consolidate light gathering and electrical vitality stockpiling inside. The charge stockpiling of NCN-PHI depends on the photograph decrease of the carbon nitride. The charge is put away by adsorption of salt metal particles inside the NCN-PHI layers. The photograph decreased carbon nitride would thus be able to be depicted as a battery anode functioning as a pseudo-capacitor, which can store light-actuated charge by catching electrons for a few hours. The achievability of light-incited electrical vitality stockpiling and delivery on request by a solitary segment light-charged battery gives an extraordinary answer for vitality stockpiling.

By evaluating writing, the philosophy of sunlight-based force is built. Sun based force age innovation is partitioned into three sections, PV innovation that utilizes the photoelectric impact to legitimately change daylight to power, concentrated sun oriented force that warms H_2O into steam to control machines, for example, power turbines, and capacity frameworks (e.g., batteries) for continuous flexibly of power when daylight isn't accessible [25]. Figure 9.5 illustrates the ideas and recently determined innovation structure recognized by extensive surveys. The philosophy blueprint, as an organized information map, is iteratively developed for the most part from writing audits (and can be refreshed by the cutting-edge patent surveys), as nitty gritty depictions of key sun powered advancements and their connections.

9.2.2 Mining of the Patents

The advances distinguished may likewise be extended to PV force age gadgets. [26] depicted the innovative advancement of PV cells utilizing licenses investigation. The outcomes showed that the PV licenses are moved in three regions: PV semiconductor materials, direct change of light vitality into electric vitality, and sun-powered boards adjusted for rooftop structures. Moreover, natural polymers, carbon nanostructures, mixes III-V, and cadmium cells are viewed as exceptional cases of PV cells licenses.

Bunching is a utilization of solo AI and partitions reports into bunches dependent on their connections [27]. A decent bunch result includes more noteworthy likeness inside a similar gathering however littler comparability between various groups. By investigating watchword terms that showed up in space licenses, licenses with comparative catchphrase terms are bunched into the gatherings. Trappey et al. [28] utilized standardized TF-IDF to locate the key terms in the corpus of 3D printing licenses, thinking about various lengths of patent archives, for progressive bunching, K-means, and K-medoids to more readily examine patent sub-innovation groups. [29] additionally suggested a way to deal with gauge, promising advances by grouping licenses. A balanced patent–patent network is developed by ascertaining the Pearson's relationship coefficient between patent reports. At that point, the k-implies calculation is utilized to bunch licenses with the normal outline width applied to decide the best number of groups [30]. The point for groups is characterized by inspecting the mix of patent arrangement classes from each bunch. At long last, patent pointers, for example, forward references, triadic patent families, and free cases, are broken down to sum up the promising advancements.

9.2.3 Latent Dirichlet Allocation

LDA can all the more likely dispose of word vagueness and allot records more precisely to subjects. Zou built up a shrewd strategy for building a metaphysics utilizing the LDA subject displaying and recognizing the key expressions under every theme from countless patent records. The patent archives are grouped utilizing the k-implies and various leveled bunching techniques, and afterward the LDA subject model is fabricated depending on each group. The quantity of themes is controlled by scientists by watching the model preparing results. After watching results, they analyzed them and develop models accordingly. The vital expressions in any theme become the yield, thereby developing the license philosophy of that particular area. The points are kept constant. The time data inside a particular model is considered as a variable. These variables are utilized to find the shrouded themes. Any change in words alongside changes in time are utilized to distinguish point designs. Doucet et al. give a methodology that permits a changeable model with respect to time. This could be utilized for consecutive significance examination or molecule sifting. Canini et al. depict an execution of online LDA structure with molecule channels that produces preferable outcomes over numerous LDA runs.

Most subjects displaying calculations that address the advancement of archives after some time utilize a similar set of points. This implies that new themes emerge, and old ones vanish. Wilson and Robinson proposed a calculation to demonstrate the birth and passing of points inside an LDA-like structure. The client initially chooses an underlying set of points. Afterward, the new subjects, i.e. set of points, can be made or resigned without oversight. The calculation of this exploration gives beginning themes. The initial step processes the float of any point as for its partner, where every subject is a likelihood circulation. This permits the utilization of the Hellinger helpful disparity measure. After registering the float for all subjects in a particular age group, it can decide whether any other sets have sufficiently changed so as to produce another theme. The altered Z score is utilized to distinguish the focal inclination of every subject and to figure out which points have floated excessively far and should be apart. Then again, old subjects are joined into a bigger conversation or dropped altogether. The strategy quantifies the quantity of tokens relegated to every theme. Points with less tokens are set waiting on the post-trial process. On the off chance that a theme remains waiting on the post-trial process for more than 10 ages, that point is set apart as shut.

9.3 RELATED WORK

The creators suggest that the business must flourish as per the industry standard. Some important discoveries, seen in other contemplates, are, for example, generous enhancements for long haul gauging when utilizing numerical climate expectations (NWP) as information. Moreover, forecasting the position of a cloud with satellite-based information has improved momentary sun-oriented PV vitality yield gauging. Thus, model exactness will, in general, differ contingent upon the climatic state of the formative area. As of this, a model is probably going to perform better in one area over another when used on different locales at the same time. Likewise, as climatic situations can oscillate throughout the yearly cycle, a particular model may provide excellent results at a particular point when prepared for a specific climate rather than many climatic conditions [7].

Usually, the primary ends are used for determining the load forecast using AI or conventional weather forecasting methods. Both these methods provide fruitful results [31]. Even for determining power costs, a wide blend of time arrangement and ML strategies has been utilized. Additionally, the article provides a blueprint of methods utilized in the opposition for windmills; and sun-based PV boards' power determination. As contrasting to stack appeal and power costs, the assortment of measured approximating methods for windmills and sun-oriented PV boards power yield is lesser. The ML procedures have been applied in few cases while it is the conventional time arrangements that find wide applications and, hence, they are largely utilized.

The flashing climate condition is probably not going to alter drastically over a short period (specifically throughout days with unwavering climate) nor it does, in this way, it generally gives a precise intermediary of things to come transient vitality yield [32]. All in all, for transient estimating, ARIMA and ARIMA along through fictitious information sources will be significant models to research just as considering the utilization of slacked sunlight-based vitality yields [8].

The creators additionally figured out how to accomplish exactness upgrades for the ANN model when utilizing a hereditary calculation (an advancement calculation roused by the regular determination process) as their streamlining calculation.

Eventually, Coimbra and Pedro recognize the changing gauging precision in various climate systems. They propose a division of information for various climate systems when displaying. Here, the creators accept that appropriate pattern to a particular climate system cluster of the data gets better and better prescient outcomes contrasted with utilizing template to a specific data for all extraordinary climate systems. The principle approaches utilized in the examination were Head Part Investigation (PCA) and a component designing system in mix with an Inclination Boosting Tree model. Besides, the creators utilized diverse smoothing ways to deal with make highlights from their NWP information. In particular, the creators utilized a matrix of NWP information around the area of the PV establishment and registered spatial midpoints and differences of climate boundaries [33]. Other than making highlights dependent on a nearby matrix of focuses, the creators likewise figured fluctuations for various indicators for distinctive lead times. While building change highlights dependent on lead times, the hidden thought was that the component would show the fluctuation of the climate [11].

The principle end is superior outcomes from utilizing mutually PCA and highlight building. As indicated by the creators, there is a twin information hole for the additional examination. The primary perspective concerns highlight the board (include designing and highlight determination) and even more solidly with respect to how to make important highlights that improve the gauge. The subsequent angle alarms the subject of advance investigating ML displaying procedures that can be actualized in blend with instructive highlights. Their last remark is that profound learning procedures will be a fascinating way to seek after blending with legitimate element the executives.

Factual techniques use rich arrangement of estimated sun-oriented creation information, sun-based irradiance information to help in constructing the framework that can be made to absorb the connection concerning climate settings, and the objective mutable of sun-oriented creation or sun-based irradiance esteems [12]. Our work looks like to the errands in the ground of audio-visual characterization and activity acknowledgment where the knowledge chore is utilized on successions of pictures establishing as a 4D tensor key information. As run-of-the-mill picture 4D key information, the measurements are shading channel, time, tallness, and thickness, in proposed effort, the elements of the 4D input information is climate highlights (14–18 highlights), time, tallness and width. Nonetheless, past chips at 3D CNN has led the scientists in blueprint of the prototype formation. In Ref. [13], the creators utilize the 3D CNN, which altogether beats the edge-based 2D CNN for most of the applications, and extra elevated echelon highlights can help the general execution in real-life acknowledgment in recordings. In Ref. [14], the creators show that step by step amalgamating deep space, period data along constructing further systems accomplishes improved outcomes. Recordings were transformed into arrangement of trajectories (picture fixes or includes) and utilized the Intermittent Neural System through a Long Momentary Memory (LSTM) mechanism also, LSTM decoder progression for the assignment of activity acknowledgment in film and next edge expectation [15]. If we talk about the implementation of the associated performance in video grouping and activity acknowledgment and the qualities of our concern, we define our concern to a great extent into three classes:

- Utilizing volumetric convolution, where a 4D climate input is indicated by the six "outline" and yield is a magnitude.
- Utilizing the volumetric convolution, where the yield is hourly based for a day, representing direction.

In Figure 9.3, it is observed that there are six parameters, namely, panel temperature, wind direction, irradiance, air temperature, precipitation, and wind speed, which determine the power of the solar cell; hence, there are six parameters that needs to be considered while finding power of the PV cells. So, $r=6$ parameters have been considered.

Hybrid machine deepen learning model has the option to anticipate few hours forward sunlight-based irradiance. Their outcomes indicated that the half-breed paradigm achieved superior outcomes than any single ML model utilized in the cross-breed model [34]. A new investigation by Koprinska et al. delineated static and dynamic ways to deal with gathering the sunlight-based force forecast of NNs [35]. Their analysis indicated overpowering outcomes for the gathering approaches contrasted with packing, boosting, arbitrary woodland, and four single forecast models (NN, SVM, K-NN, and a diligence model). Constrained exploration can be found on troupe measurable models for unrivaled execution and exactness. The creators in Ref. [36] investigated eight group strategies to consolidate the aftereffects of six best models from every group of factual models: SARIMA (34 models), ETS (28 models), MLP (2 models), STL deterioration (4 models), TBATS (70 models), and the Theta model (2 models). Even though outfit discovering demonstrated excessive effective execution and exact outcomes in numerous documents, the outcomes in Ref. [36] indicated minor improvement from the superlative paradigm: the explanation being that the models tried brought about profoundly corresponded blunders. This drove us to underline that assorted variety is the key to significant improvement of group techniques. The creators in Ref. [37] introduced a bunch-based methodology applied to the worldwide sun powered radiation. Their methodology at that point foresaw the even worldwide sun-oriented radiation utilizing a blend of two ML models: SVM and ANN. The outcomes indicated elevated anticipating exactness contrasted with the traditional ANN and SVM.

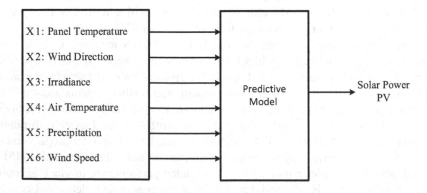

FIGURE 9.3 Predictive model of the solar PV power.

9.4 MATHEMATICAL TERM TECHNIQUES

Techniques that have been used and implemented to reduce the errors and help get better solar power during the forecasting and rainy weather, as it is somewhat complicated to detect the correct power.

9.4.1 MOVING AVERAGE PROCESS

Form of the regression which is being used on the external asymmetrical shocks and the prediction parameters dependent on it, it does not work on the historical data. K_j will be independent on the values of j which will act as threshold values of candidate generation pruned value. For $j = 0$, 0.25, and 0.5, the value of the $\Theta = 1$, and this will not be the perfect value for the power detection during overcast weather [37].

9.4.2 ARIMA

Autoregressive integrated moving average helps in the differencing which has been performed on K_t where $K_t K_1$ rather than taking the K_0 into the consideration. When $j = 0$, the model is coined as ARIMA and the stationarity is attained and reduces the mean time dependence [16].

9.4.3 SARIMA

In this process, the seasonal pattern will be recognized with the binary values that have been added and mapped. The resultant pattern will be in the form of the daily, monthly, and yearly basis. Few attribute variables are there that signify A, B, and D having similar properties as a, b, and d with the step of the seasonal change. If $B = 1$, then the differential seasonal is expressed as follows:

$$(1-K)^{1xr} Q_t = Q_t - Q_{t-r} \qquad (9.1)$$

This is to take into the consideration that we are working with pictorial representation of the seasonal data whose threshold frequency should not be highly fluctuating. It should be constant, so that it should not be autocorrelated during different processes of regression.

9.4.4 K NEAREST NEIGHBORS

The algorithm is being used to estimate the situational dissemination of "Z" and "W" and later assigning Z to the class of the highest probability. Then the Euclidean distance formula is used to measure the closest point by using the training data.

When "K" is growing, the boundary of the decision starts coming closer to the linear which states the high bias. To choose the correct "K" which impacts the tradeoff bias – variance and using the cross validation. This is to note that for the regression used data the implementation of the KNN will be intuitively. So, the correct value of K can be determining for some near metric distance and then be split by the average value.

9.4.5 REGRESSION TREES

Here we are tending to minimize the total of the square residuals for each split block to find the parameter called predictor parameter that reduces the RSS at every split. Then it will be recursively reiterated for individual subregion that is generated by Kth repetitions. The "K" can be defined as the number of the various regions where the entire space is partitioned. We need to define the total number of the regions so it's been supervised learning.

9.5 DESIGNING

Below are the different parameters that are taken into the consideration, which helps in designing.

9.5.1 LONG SHORT-TERM MEMORIES

The long transient memory systems (LSTM), as shown in Figure 9.4, are exceptional intermittent neural systems (RNNs) which were formerly presented in Ref. [37]. An RNN is a neural system (NN) with repetitive associations among the various neurons that empowers it to gain from the present and the past data to locate a superior arrangement. Be that as it may, when two cells in an RNN are far off from one another, then it becomes hard to get helpful data because of the slope evaporating and blast issues. Still, there is possibility to get information through specific neurons termed as memory cells. These uncommon neurons are utilized in LSTM so that they can store valuable data over a subjective timeframe.

The figure represents the cells in the LSTM having the ability to train the data which can be readable, stored, and erased as per the need to adjust the enamours controlling gates. Here, $C[t-1]$ and $h[t-1]$ are the previous parameters, while $x[t]$, $h[t]$, and $c[t]$ are present parameters.

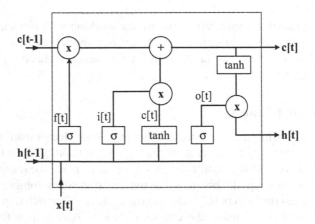

FIGURE 9.4 Flow of the LSTM.

9.5.2 Auto STM

An AE is an unaided neural system where the info and the yield layers have a similar magnitude. It attempts to become familiar with the character work so the info "x" is roughly like the yield with certain imperatives applied to the system, e.g., a predetermined quantity of neurons in the shrouded layer contrasted with the information layer. In Figure 9.5, the various attributes that are used to estimate power is shown.

In this manner, an AE goes about as a blower and a decompressor comprising of two sections isolated by a blockage at the middle:

- encoding side, where the neurons are diminished from the info layer to the concealed layer, and
- interpreting side, where the layers in the encoding side are reflected.

9.5.3 Statistical Forecasting

The latest time arrangement lines are called lines of the theta and keep up the mean and the incline of the first run-through arrangement. In addition, the collapses of the new time arrangement ebbs and flows rely upon the estimation of Theta coefficient, i.e., to distinguish the drawn-out practices of the time arrangement dataset programmed the Theta coefficient to be somewhere in the range of 0 and 1 ($0 < 1$). In any case, when >1, the new theta line is more expanded, it influences the transient patterns. The theta lines are then extrapolated independently and consolidated to produce the anticipated sun-oriented PV power. The creators in Ref. [17] decayed the first run-through arrangement into two theta lines by setting the theta coefficient to 0 and 2. The main line ($k (= 0)$) speaks to the straight relapse line of the first run-through arrangement amplifying the drawn out patterns. The subsequent line ($k (= 2)$) copies the first arch, amplifying the transient patterns. In this, the gauging

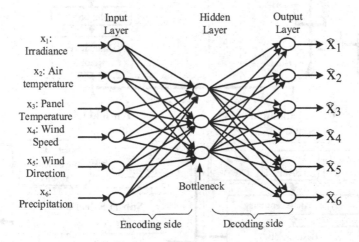

FIGURE 9.5 Attributes used to evaluate the power.

procedure is practiced by directly extrapolating the primary theta-line while extrapolating the subsequent line utilizing basic exponential smoothing (SES). A short time later, the determined time arrangement of the two theta-lines is basically consolidated through equivalent loads bringing about the last conjecture of a particular time arrangement dataset.

9.5.4 Flame Working

The system comprises five stages: information assortment, information pre-processing, cross-approval (CV), ten times cross-approval and subset determination as indicated in Figure 9.6. The information assortment step accumulates the data from four information sources, sunlight-based force age information, sun-oriented rise, observational, and gauge climate information.

Sunlight-based force age information is given by the Korean Open Information Entry (http://www.data.go.kr). Sun-based rise information is given by the planetarium programming Stellarium (http://www.stellarium.org) and observational and conjecture climate information are offered by the Korea Meteorological Organization

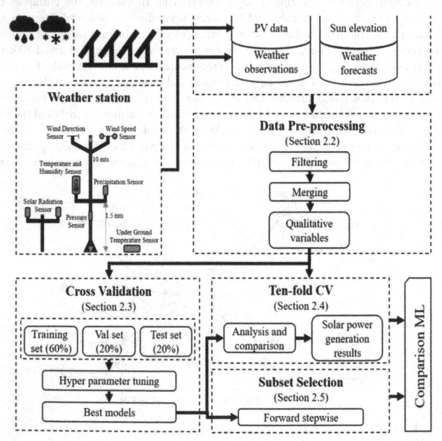

FIGURE 9.6   Workflow of the extracting threshold.

(KMA) (http://kma.go.kr). The sun-oriented force age information utilized in this exploration is given by the Korean government from Yeongam in South Korea. The sun-based force information is present in 1-hour time frames from 1st January 2013 to 31st December 2015.

Alongside the sun-oriented force age information, we likewise gather three extra information that will be utilized as the contributions of AI models. Right off the bat, the sun-oriented height information is taken from the topographical area of the force plant in Yeongam. The sun-oriented height speaks to the situation of the sun from the force plant viewpoint in scope and longitude. Next, the observational climate records gave by KMA are the deliberate real climate. The observational climate information is taken from the nearest meteorological station situated in Mokpo, roughly 8.5 km from the PV plant. At long last, the gauge climate records reported by KMA are allowed in 3-hour time frames beginning from 02:00 a.m. consistently. In this examination, the figure climate information utilized are the expectations 36 h-ahead.

The meteorological perception framework used by KMA comprises a 10-m-high meteorological pinnacle. At the top, wind course and wind speed sensors are introduced on a level plane on the left and right sides. The moistness and temperature meter are introduced at 1.5 m over the ground and a precipitation sensor is introduced on the opposite side. A weight sensor is introduced around 50–60 cm over the ground.

9.6 MODEL VIEWS

There are many model views available in the literature. Three of them have been applied in this paper; hence, they are only discussed here.

9.6.1 DEEP LEARNING

In research paper, three different learning plans have been discussed that deals the fleeting development of the climate data also, analyse its impact on sun-oriented vitality yield. The study is based on a particular time frame on each day basis. Its starts from 5.00 a.m. in the morning and ends at 9.00 p.m. in the night. So, the net hour of study in a day is 17 hours. The recurrence of target sun-based vitality yield is hourly; in this way there are 17 objective qualities day by day too. For every day, there are different choices to expand the availability of the system in time measurement to become familiar with the spatio-fleeting highlights of the climate information. The network on fleeting measurement should be possible at various stages. In this paper, different things have been tried with three fleeting network designs: single-hour model (SHM), every day model, half-breed model [38]. The SHM just considers fractional hour transient networks with a four-dimensional tensor comprising of climate framework information. The day-by-day model, likewise, takes the 4D tensor comprising the climate data for the 17 hours and yields the related 17 hourly sunlight-based yield mutually. Finally, the cross-breed model uses shared convolution organization for each single hour input, at that point the yield from the convolution organized for every individual hour is the contribution to a repetitive neural system (LSTM utilized in this paper), and the RNN will yield the 17 scores of the hourly sun-oriented vitality yield expectation.

9.6.2 DAILY VERTEBO

When we are approaching to the daily model which is dependent on the 4D tensor, we can notice that we will be taking the sample of 6 with the 10-minute frequency and we are having the 17 hours which leads the sample quantity of $17 \times 6 = 102$. There is significant reason for using the 17 hours instead of 24 hours because there is no sunlight at night; so, the output energy $= 0$. Hence, we do not need to account for it.

The complete design for the Alexnet centered quotidian prototype is:

K D(128, 2, 11, 11; 1, 4, 4) → KP(1, 3, 3; 1, 2, 2) →
K D(384, 2, 5, 5; 1, 1, 1) → KP(1, 3, 3; 1, 2, 2) →
K D(768, 2, 3, 3; 1, 1, 1) →
K D(576, 1, 3, 3; 1, 1, 1) →
K D(576, 1, 3, 3; 1, 1, 1) → KP(1, 3, 3; 1, 2, 2) →

The FC value will be 16 here because 8 p.m.–9 p.m. value $= 0$.

9.7 DATA PROCESSING

The processing of data depends on temporal weather grid, mutability, and data mining with visualization, which are discussed in the following.

9.7.1 TEMPORAL WEATHER GRID

A transient network and worldly inconstancy for the highlights are registered. As the NWP factors are estimated for the Postal district, and not the specific directions of the particular site, one can expect that the 10-minutes slacked, and lead gauges have prescient capacity. At that point, if $NWP_{r,k}$ is the figure of NWP alterable "r" at time "k", the slacked esteem is $NWP_{r,k+1}$ and the lead esteem $NWP_{r,k+1}$ where each time-step is 10 minutes. The framework is just performed for $k \pm 1$, not to lose a lot of information.

9.7.2 MUTABILITY

The inconstancy of the NWP was figured by utilizing the 10-minute slacked esteem, the 10-minute control esteem, and the genuine estimation of the specific climate boundary. The thought was to make an irregular that catches likely changeability in the climate. For model, high fluctuation in viable overcast spread is probably going to influence the vitality yield, making it a possibly significant information.

9.7.3 DATA MINING WITH VISUALIZATION

R Studios with mining are utilized to produce solo learning outcomes. The R bundles utilized are depicted in this section. PD directory is a bundle to change information with the goal that it is anything but difficult to deal with, including the information casing and arrangement where information can be taken apart and joined. Excel files are utilized to peruse Exceed expectations records by filename and the RNK oversee text pre-processing, for example, decrease, tokenization, and stop word or accentuation evacuation. NP abbreviation is broadly used to store information into exhibits for scientific tasks. Genism has numerous capacities for text mining and characteristic language preparation. Linear discrimination analysis is utilized in this exploration. Scikit is utilized for information mining and information investigation including transaction id, cosine comparability and K-implies.

The innovation advancement among licenses and scholarly writing is very extraordinary. As recently expressed, most licenses depict sun-oriented hydropower stockpiling frameworks with an assortment of subsystems applicable to circuitous sun-based assortment innovation. Scholarly writing, most as often as possible, proposes new structures or calculations for framework-associated power gracefully frameworks. Additional framework structures must be painstakingly gotten ready for exact execution and combination since new methodologies require complex assessment forms and different factors, for example, social and government acknowledgment. Frameworks are generally simple to actualize in the event that they are enhancements dependent on existing frameworks. On the other hand, it is hard to execute inventive frameworks and calculations as a first endeavour. Along these lines, it is sensible that there are more novel advances portraying the coordination of sustainable power source age frameworks and recreation of network-associated vitality stockpiling frameworks in the writing, while innovations depicting sunlight-based hydropower stockpiling frameworks are introduced in licenses.

The standard innovation develops from off-framework to lattice-associated frameworks including advancements, for example, self-balance stockpiling frameworks and remote-coordinated sensors which are basic for shrewd matrix systems. The development chart shows that sun-powered innovation is drifting toward wise vitality gracefully frameworks. The keen lattice power flexibly framework can be incorporated with digital material science frameworks and the sustainable assets industry. Keen battery the executives and gracefully balance frameworks are fundamental pieces of the digital physical framework.

9.8 CROSS-VALIDATION

The capriciousness of the NWP was figured by using the 10-minute loosened regard, the 10-minute lead regard, and the authentic estimation of the particular atmosphere limit. The thought was to make a variable that gets likely flexibility in the atmosphere. For model, high change in reasonable cloudy spread is most probably going to impact the essentialness yield, making it a conceivably important of data.

Figure 9.7 demonstrates the procedure if the information for a site was from February 2014 to the furthest limit of February 2018. The essential supposition that will be that every half-year overlap catches a pattern of occasional fluctuation, implying that vitality yield of January-June and July-December roughly follows a similar example. Every half-year overlay can, in this manner, be viewed as practically autonomous of each other. The sequential relationship that makes time arrangement dangerous for ordinary CV is just kept up inside the half-year pleats. As of this, the pleats cannot be separated arbitrarily yet rather in a progressive manner. In a perfect

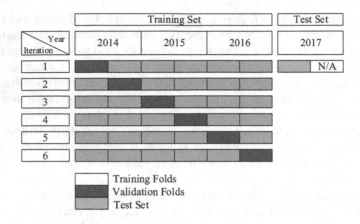

FIGURE 9.7   Cross validation results.

world, one would need to utilize an entire 1-year cycle per overlay; but the absence of adequate information restricted this.

9.9   ISSUES STILL IN WORK

There are number of issues which need to be sorted out. They are discussed in what follows.

9.9.1   METRICS ERROR

As perceptions in the first part of the day and evening for the most part have lower yield by and large, huge numbers of the supreme forecast mistakes during those occasions are lower contrasted with those during the pinnacle hours, bringing about a supported RMSE as a normal is registered. When the figure skyline is 5 hours, it ought to be noticed that most pinnacle hour perceptions are expelled from the dataset because of accessibility, and this may support the outcomes. In a business setting, it would be important to alter the mistake measurements for the pinnacle hours, since they are progressively basic for the business. In any case, no past examination that we have seen has utilized a balanced mistake measure; along these lines, executing this would make it complex to contrast our expectations and different investigations.

9.9.2   METHODOLOGY

The model developed is applied to study the various seasons taking place through the year, especially wintry weather and late spring where climatic condition drastically changes. Hence, considering these parts of the season will help to optimize the simulation so as to achieve best outcome for developing a particular model. Thus, this pattern will perform better when prepared on increasingly explicit periods, one could attempt to display just during peak hours. This helps to use a single model throughout the year as the worse climatic condition has been used to evaluate performance of the model which could be easily extended to other seasons. In the event that one would prepare the model for just winter months, the model could exclusively concentrate on these focuses without doing a trade-off between fitting the mid-year information well and fitting the winter information well. On the other hand, having various models relying upon the atmosphere would mean having to change models every now and then, which might be dull in an operational setting.

9.10   RESULTS COMPARISON

The most reduced MSE for Poly-SI is 0.0197 utilizing EN4 for T-MLSHM which is 18.93% better than GRU model as appeared in Table 9.2. Besides, the force expectation for TSCF ranch achieved an MSE of 0.00185 utilizing EN4 for T-MLSHM, a 36.21% improvement over GRU model. Table 9.3 featured the least MSE for Cocoa Single Poly-SI dataset, which arrived at 0.0168 for EN4 of T-MLSHM that brought about a 4.55% decrease of blunder rate and improved execution. In this way, T-MLSTHM utilizing EN4 was the most precise group strategy since it expanded the exactness of sun powered force forecast somewhere in the range of 4.55% and 36.21% over the single conventional models. Then again, Auto-MLSHM utilizing EN4 expanded the precision of the sunlight-based force expectation run somewhere in the range of 8.15% and 28.90% contrasted with Auto-GRU model. One explanation behind this is EN4 conveys the loads productively where the most precise model has the best weight esteem and is not close in an incentive to different loads. From the single forecast models, GRU accomplished the best exactness over other conventional ML models and the Theta factual model.

9.11   CONCLUSION

In this work, we have looked at time arrangement procedures and AI methods for sun-based vitality anticipating across five unique destinations in Sweden. Interestingly, AI strategies were increasingly clear to execute. This investigation has looked at the changed models on an overall level. For additional research, we propose keeping contrasting distinctive AI strategies in profundity while utilizing highlight designing methodologies of numerical climate expectations.

Coordinating huge scope PV plants into the force network presents extensive issues and difficulties to the electric administrators, as it makes unsteadiness the electric lattice making the electrical administrators balance the electrical utilization and force age so as to maintain a strategic distance from misuse of vitality. In this manner, a precise sun-powered force figure is an essential prerequisite toward the eventual fate of sustainable power source plants. In this examination, we proposed a crossover model (MLSHM) that joins the forecast aftereffects of both ML models and factual technique. For our investigation we built up another ML model, Auto-GRU, that gains from authentic time arrangement information to foresee the ideal sun-based PV power. So as to support the cross-breed model, two assorted variety methods were led in this examination, i.e., basic decent variety between the outfit

TABLE 9.2
Results of Poly-Si Dataset

Approach	Theta Model	LSTM	GRU	Auto-MLSHM						T-MI	
				Auto-LSTM	Auto-GRU	EN1	EN2	EN3	EN4	EN1	EN2
nMAE	0.057	0.0536	0.0346	0.0806	0.0526	0.044	0.0438	0.0438	0.0424	0.0341	0.034
nMSE	0.00695	0.0037	0.00243	0.00891	0.00429	0.00318	0.00316	0.00316	0.00305	0.00213	0.002

TABLE 9.3
Results of TSCF Data

Approach	Theta Model	LSTM	GRU	Auto-MLSHM						T-MI	
				Auto- LSTM	Auto- GRU	EN1	EN2	EN3	EN4	EN1	EN2
nMAE	0.0574	0.0698	0.0358	0.0831	0.0394	0.0411	0.0409	0.0409	0.0389	0.0354	0.0352
nMSE	0.00656	0.00577	0.0029	0.00961	0.00251	0.00303	0.003	0.003	0.00265	0.00209	0.00207

individuals and information decent variety between the preparation sets of the ML models. Four distinctive blend techniques show to join the expectation of ML models and factual strategy.

An assortment of profound learning models for the undertaking of sun-oriented vitality yield conjecture. To our best knowledge, this is the first occasion when that profound learning has been utilized on the undertaking of sun-oriented vitality yield expectation. Our model exploits the rich information produced by meteorological reproduction of climate figure. We led broad examinations on different model structures including convolutional arrangements and repetitive systems and provided improvement plans. We utilized the business acknowledged mistake estimation of the limit standardized Mean Outright Rate Blunder (rMAP E) to measure the parameters of the models. It is discovered that the hourly AlexNet organize plays out the best with an rM gorilla of 11.8%, which is at an exceptional exactness level looking at accessible literary works and our component-designed SVR model. The comparability highlights, successful in SVR model, doesn't bring improvement to the profound learning model. Our work shows that in the assignment of sun-oriented estimating, profound learning can beat advanced physical models and human-include designed models.

It may be concluded that the deep learning approach is not cent percent accurate but they provide the solution towards the faded weather where frequency of solar radiation is not constant or the weather various is more.

REFERENCES

1. Solar Power Forecasting with Machine Learning Techniques. *Science Direct – Applied Computing and Informatics*, November 2019.
2. IRENA. Renewable power generation costs in 2017. Technical report, International Renewable Energy Agency, Abu Dhabi, January 2018.
3. Jose R. Andrade and Ricardo J. Bessa, Improving renewable energy forecasting with a grid of numerical weather predictions. *IEEE Transactions on Sustainable Energy*, 8(4):1571–1580, October 2017.
4. A. Tuohy, J. Zack, S. E. Haupt, J. Sharp, M. Ahlstrom, S. Dise, E. Grimit, C. Mohrlen, M. Lange, M. G. Casado, J. Black, M. Marquis, and C. Collier, Solar fore casting. *IEEE Power & Energy Magazine*, November/December 2015.
5. R. Perez, S. Kivalov, J. Schlemmer, K. Hemker, D. Renne, and T. E. Hoff, Validation of short and medium term operational solar radiation forecasts in the US. *Solar Energy*, 84(12):2161–2172, 2010.
6. R. Zhang, M. Feng, W. Zhang, S. Lu, and F. Wang, Forecast of solar energy production - A Deep learning approach. *2018 IEEE International Conference on Big Knowledge (ICBK)*, Singapore, 2018, pp. 73–82, doi: 10.1109/ICBK.2018.00018.
7. J. Shi, W. J. Lee, Y. Liu, Y. Yang, and P. Wang. Forecasting power output of photovoltaic systems based on weather classification and support vector machines. *IEEE Transactions on Industry Applications*, 48(3):1064–1069, May 2012.
8. Tao Hong, Pierre Pinson, Shu Fan, Hamidreza Zareipour, Alberto Troccoli, and Rob J. Hyndman. Probabilistic energy forecasting: Global energy forecasting competition 2014 and beyond. *International Journal of Forecasting*, 32(3):896–913, 2016.
9. P. G. V. Sampaio, M. O. A. González, R. M. de Vasconcelos, M. A. T. dos Santos, J. C. de Toledo, and J. P. P. Pereira, Photovoltaic technologies: Mapping from patent analysis. *Renewable and Sustainable Energy Reviews*, 93:215–224, 2018.

10. T. Maeda, H. Ito, Y. Hasegawa, Z. Zhou, and M. Ishida, Study on control method of the stand-alone direct-coupling photovoltaic–Water electrolyzes. *International Journal of Hydrogen Energy*, 37:4819–4828, 2012.

11. D. Randall Wilson and Tony R. Martinez. The general inefficiency of batch training for gradient descent learning. *Neural Networks*, 16(10):1429–1451, 2003.

12. R. Zavadil, Renewable generation forecasting: The science, applications, and outlook. In *System Sciences (HICSS), 2013 46th Hawaii International Conference on*. IEEE, 2013, pp. 2252–2260.

13. A. Mellit and A. M. Pavan, A 24-h forecast of solar irradiance using artificial neural network: Application for performance prediction of a grid-connected PV plant at Trieste, Italy. *Solar Energy*, 84(5):807–821, 2010.

14. B. Goswami and G. Bhandari, Automatically adjusting cloud movement prediction model from satellite infrared images. In *India Conference (INDICON), 2011 Annual IEEE*. IEEE, 2011, pp. 1–4.

15. A. Radovan and Z. Ban, Predictions of cloud movements and the sun cover duration. In *Information and Communication Technology, Electronics and Microelectronics (MIPRO), 2014 37th International Convention on*. IEEE, 2014, pp. 1210–1215.

16. L. Gigoni, A. Betti, E. Crisostomi, A. Franco, M. Tucci, F. Bizzarri, and D. Mucci, Day-ahead hourly forecasting of power generation from photovoltaic plants. *IEEE Transactions on Sustainable Energy*, 9(2):831–842, 2018.

17. M. Torabi, A. Mosavi, P. Ozturk, A. Varkonyi-Koczy, and V. Istvan, A hybrid machine learning approach for daily prediction of solar radiation. *International Conference on Global Research and Education*, Springer, 2018, pp. 266–274.

18. M. E. Glavin, P. K. Chan, S. Armstrong, and W. G. Hurley, A stand-alone photovoltaic supercapacitor battery hybrid energy storage system. In *Proceedings of the 2008 13th Power Electronics and Motion Control Conference (EPE-PEMC)*, Poznan, Poland, 1–3 September 2008, pp. 1688–1695.

19. B. Marion, A. Anderberg, C. Deline, J. del Cueto, M. Muller, G. Perrin, et al., New data set for validating PV module performance models. In *Photovoltaic Specialist Conference (PVSC), 2014 IEEE 40th*. IEEE, 2014, pp. 1362–1366.

20. B. Goswami and G. Bhandari, Automatically adjusting cloud movement prediction model from satellite infrared images. In *India Conference (INDICON), 2011 Annual IEEE*. IEEE, 2011, pp. 1–4.

21. A. Zahedi, Maximizing solar PV energy penetration using energy storage technology. *Renewable and Sustainable Energy Reviews*, 15:866–870, 2011.

22. C. H. Zou, Using Non-Supervised Machine Learning Approach to Generate Knowledge Ontology for Patent (Advisor: A.J.C. Trappey). Master's Thesis, Department of Industrial Engineering and Engineering Management, National Tsing Hua University, Hsinchu, Taiwan, 2018.

23. X. Wang and A. McCallum, Topics over time: A non-Markov continuous-time model of topical trends. In *Proceedings of the 12th ACM SIGKDD International Conference on Knowledge Discovery and Data Mining*, Philadelphia, PA, USA, 20–23 August 2006, pp. 424–433.

24. J. H. Teng, S. W. Luan, D. J. Lee, and Y. Q. Huang, Optimal charging/discharging scheduling of battery storage systems for distribution systems interconnected with sizeable PV generation systems. *IEEE Transactions on Power Systems*, 28:1425–1433, 2013.

25. H. Sridhar and K.S. Meera, Study of grid connected solar photovoltaic system using real time digital simulator. In *Proceedings of the 2014 International Conference on Advances in Electronics Computers and Communications*, Bangalore, India, 10–11 October 2014, pp. 1–6.

26. D. Yang, X. Wang, and J. Kang, SWOT Analysis of the Development of Green Energy Industry in China: Taking solar energy industry as an example. In *Proceedings of the 2018 2nd International Conference on Green Energy and Applications (ICGEA)*, Singapore, 24–26 March 2018, pp. 103–107.

27. G. Hinton, L. Deng, D. Yu, G. E. Dahl, A.-r. Mohamed, N. Jaitly, et al., Deep neural networks for acoustic modeling in speech recognition: The shared views of four research groups. *IEEE Signal Processing Magazine*, 29(6):82–97, 2012.

28. A. J. C. Trappey, C. V. Trappey, C.-Y. Fan, A. P. T. Hsu, X. K. Li, and I. J. Y. Lee, IoT patent roadmap for smart logistic service provision in the context of Industry 4.0. *Journal of the Chinese Institute Engineers*, 40:593–602, 2017.

29. Z. S. Li, G. Q. Zhang, D. M. Li, J. Zhou, L. J. Li, and L. X. Li, Application and development of solar energy in building industry and its prospects in China. *Energy Policy*, 35:4121–4127, 2007.

30. X. Li, Q. Xie, J. Jiang, Y. Zhou, L. Huang, Identifying and monitoring the development trends of emerging technologies using patent analysis and Twitter data mining: The case of perovskite solar cell technology. *Technological Forecasting Social Change*, 146:687–705, 2018.

31. B. Gentry, Holographic Optical Elements. HARLIE. NASA. Archived from the original on 15 February 2013. Retrieved 9 August 2018. Available online:harlie.gsfc.nasa.gov (accessed on 9 August 2018).

32. A. J. C. Trappey, C. V. Trappey, D. Y. Wang, S. J. Li, and J. J. Ou, Evaluating renewable energy policies using hybrid clustering and analytic hierarchy process modeling. In *Proceedings of the 2014 IEEE 18th International Conference on Computer Supported Cooperative Work in Design (CSCWD)*, Hsinchu, Taiwan, 21–23 May 2014, pp. 716–720.

33. J. Appen, T. Stetz, M. Braun, and A. Schmiegel, Local voltage control strategies for PV storage systems in distribution grids. *IEEE Transactions on Smart Grid*, 5:1002–1009, 2014.

34. S. K. Kim, J. H. Jeon, C. H. Cho, J. B. Ahn, and S. H. Kwon, Dynamic modeling and control of a grid-connected hybrid generation system with versatile power transfer. *IEEE Transactions on Industrial Electronics*, 55:1677–1688, 2008.

35. Z. W. I. Koprinska, I. Koprinska, A. Troncoso, and F. Martínez-Álvarez, Static and dynamic ensembles of neural networks for solar power forecasting. *2018 International Joint Conference on Neural Networks (IJCNN)*. IEEE, 2018, pp. 1–8.

36. D. Yang and Z. Dong, Operational photovoltaics power forecasting using seasonal time series ensemble. *Solar Energy*, 166(2018):529–541, 2018.

37. A. Alzahrani, P. Shamsi, M. Ferdowsi, and C. Dagli, Solar irradiance forecasting using deep recurrent neural networks Renewable Energy Research and Applications (ICRERA). *2017 IEEE 6th International Conference on*. IEEE, 2017, pp. 988–994.

38. S. K. Kim, J. H. Jeon, C. H. Cho, J. B. Ahn, and S. H. Kwon, Dynamic modeling and control of a grid-connected hybrid generation system with versatile power transfer. *IEEE Transctions on Industrial Electronics*, 55:1677–1688, 2008.

10 The Imperative Role of Solar Power Assistance for Embedded Based Climatic Parameters Measurement Systems

Safia A. Kazmi
ZHCET, Aligarh Muslim University, Aligarh, India

Ankur Kumar Gupta
IIMT University, Meerut, India

Nafees Uddin
JIMS Engineering Management Technical
Campus, Greater Noida, India

Yogesh K. Chauhan
Kamla Nehru Institute of Technology, Sultanpur, India

CONTENTS

10.1 INTRODUCTION

Today, solar energy is gaining popularity and replacing conventional energy sources, e.g. coal, oil, and gas, for electricity generation. The process of generation of electricity from fossil fuels leads to air, water, and land pollution, which affects the forests and environment. The advantage of using solar energy is that it can be utilized directly for a stand-alone low-scale power system and can be used by the end consumers as well as for commercial applications. The distributed power generation nature of solar-powered systems makes it a major and viable renewable energy source in rural areas and in roof-top installations in urban areas. Various technologies and applications are adopted to harness the solar energy, such as the solar panels, solar heaters, and the solar architecture for power generation in household and industrial applications. It is generally used to heat homes, provide lighting, heat water, and generate electricity. Basically, the proper usage of energy can lead to a beautiful and bright future.

In spite of being the most abundantly available energy source, solar energy has many drawbacks, the primary one being the cost of installation as compared to other RE sources. Setting up of a solar panel yard with the semiconductor material used in the panels for direct conversion from solar to electrical energy incurs high cost. The production of semiconductors require high standards of purity and hygiene, which further escalate the overall cost involved. Another major drawback of using solar energy is the inconsistency in the availability of sunlight all through the year. This dependability on weather conditions along with high cost are the major challenges faced in the design and implementation of solar power-based monitoring systems.

The use of embedded systems, e.g. analog and digital sensors, Arduino/microcontroller, and wireless data transceiver systems, for performance enhancement of the solar PV-based applications is discussed. The basic advantages of these embedded system components are easy monitoring of performance parameters, accuracy, and measured data transmission of solar powered applications.

10.2 LITERATURE REVIEW

Due to several instances of solar PV applications, researchers explored novel approaches for monitoring of various environmental parameters. Comprehensive studies have been carried out related to the accuracy, robustness, efficiency, and execution [1–29].

In Ref. [1], the authors designed a system to monitor the solar irradiance for a one-year duration. The temperature and current sensors are utilized to measure the performance of solar PV system. A portable data acquisition system (DAS) with the Lab-VIEW-based graphical user interface (GUI) is developed and shown in Figure 10.1.

In Ref. [2], the authors developed a hardware model, which comprised sensors and data storage as primary components for the measured performance parameters. The role of the proposed system is monitoring the performance of solar PV systems. The Lab-VIEW-based GUI system is developed for real-time monitoring the $I–V$ curves. The operating duration of this developed model is 2 years, and the layout of the discussed system is shown in Figure 10.2.

The authors of Ref. [3] designed a commercial model of a data logger to measure the environmental parameters such as solar irradiance, temperature, and wind speed for 3 years. Hourly PV efficiency as a function of hourly in-plane irradiation was observed. The implemented system is shown in Figure 10.3.

The authors in Ref. [4] monitored the microscale grid-connected PV system up to 5 kWp range. For efficient and smart monitoring, the system performance was monitored for failure conditions by the satellite system. Under this investigation, an

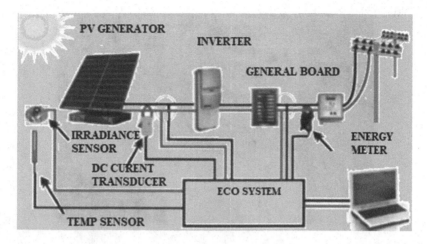

FIGURE 10.1 Block diagram showing the devices interconnected with the PV plant for performance monitoring [1].

8-month testing phase was carried out on hundred PV systems in three European countries. A stand-alone PV system is monitored by the embedded web server-based system for remote condition monitoring. The main components of the system, i.e. the temperature sensor, the voltage sensors, and the PIC18 microcontroller unit, are integrated with the local area network [5].

In Ref. [6], the authors developed a wireless-based prototype model for solar power performance monitoring system under non-ideal conditions. A Lab-VIEW-based GUI system was used for remote monitoring of electrical performance parameters such as voltage, current, and power of PV modules. The measured performance

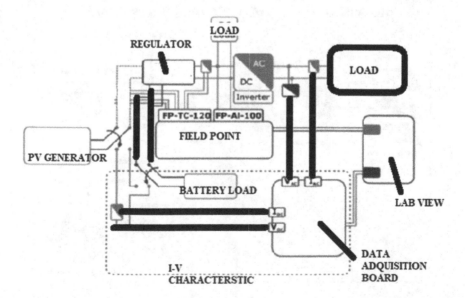

FIGURE 10.2 Layout of proposed system for performance monitoring of PV solar plant [2].

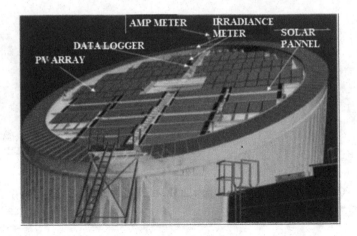

FIGURE 10.3 The PV array and monitoring equipment [3].

parameters are transferred by the ZigBee transmitting module to the operator end. The developed model is shown in Figure 10.4.

Other authors [17] have developed an advanced data logger for recording the performance parameters. For the performance improvement of solar PV system and maximum power point tracking module is used to supply voltage the wireless module JN5139. The intermediate stages are battery charger and designed model of DC–DC buck converter [7]. The developed model is shown in Figure 10.5.

The authors of Ref. [8] installed a solar PV plant of 1.28 kWp capacity. A hardware implementation of a data logger system was done for the storage of the climatic

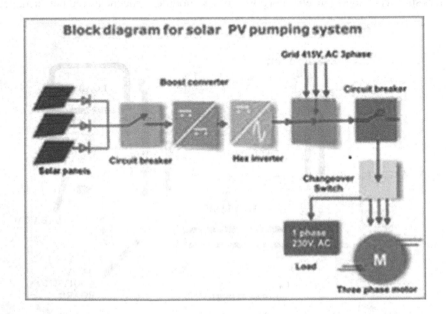

FIGURE 10.4 Performance monitoring of two PV modules using ZigBee [6].

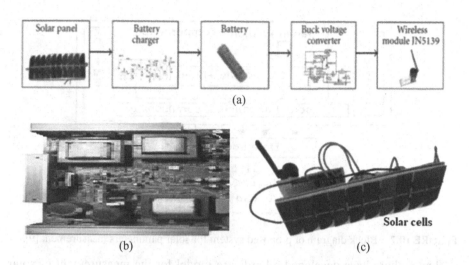

(a)

(b)   (c)

FIGURE 10.5 (a) Functional schematic diagram of the alternative power supply unit. (b) Front view: Printed circuit board and wireless module. (c) Rear-side view: solar PV cells for power backup [7].

parameters data (solar irradiation and wind speed sensor) and the electrical performance parameters of solar PV system (voltage and current). Moreover, the performance assessment of solar PV system was carried out efficiently with the main components such as microcontroller unit 30F3013 and RF transmitting modules at the receiver and transmitting ends. The proposed scheme of wireless remote monitoring system is shown in Figure 10.6.

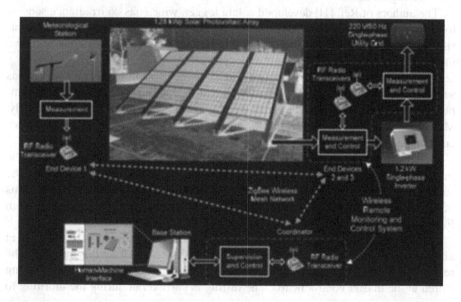

FIGURE 10.6 Wireless remote monitoring of climatic and performance parameters of a low-scale PV system [8].

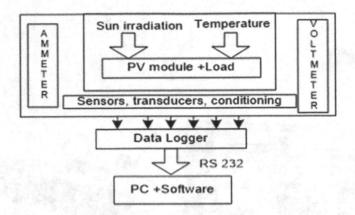

FIGURE 10.7   Block diagram of proposed system for solar parameters measurement [9].

The authors have developed a hardware model for the measurement of solar performance and environmental parameters, e.g. current, voltage, temperature, and solar irradiation level. The photovoltaic system of 120 Wp capacity is considered for the performance monitoring [9]. The developed model is shown in Figure 10.7.

In Ref. [10], wired and wireless sensor network technologies are used to develop a system for monitoring the solar PV system performance, such as voltage, current, and power under two separate irradiation conditions. Furthermore, the environmental parameters such as temperature and humidity are also monitored. The testing phase of the developed system is investigated at the laboratory scale with the ZigBee transmitting module.

The authors of Ref. [11] developed a data logger, which has 25 irradiance sensors to estimate the solar irradiation on site. Using this network of sensors, two different-sized power plants are considered in this study. The deployed sensor node is shown in Figure 10.8.

In Ref. [12], the authors developed a data logger system comprised of various sensors to assess the data at the PV system site. Numerous analog and digital temperature sensors are used for measuring cell and ambient temperature continuously. Moreover, current sensors are used for measurement of PV system current, battery current, and load current. The developed system is shown in Figure 10.9a and b.

In Ref. [13], a PLC modem-based power management system is developed for the investigation of PV systems in smart home application. PLC modems measure the status of PV modules and the user can then check the status of PV system in terms of failure conditions. The schematic diagram of the developed model is shown in Figure 10.10.

In Ref. [14], the authors placed 16 temperature sensors to measure the real data at an energy site. An embedded system called DAS was designed for this purpose. The recorded data is shown on a liquid crystal display (LCD) regularly as well as being stored at the master control board. The testing is carried out during the morning to

FIGURE 10.8 Deployed solar irradiance sensor [11].

evening hours. It is observed that the temperature is high between 9:00 AM and 3:00 PM. The block diagram of the developed model is shown in Figure 10.11.

The authors of Ref. [15] designed a low-cost power line communication (PLC)-based user-friendly PV monitoring system. To reduce the system cost, the PLC module is utilized without a communication modem. Individual PLC modules are deployed at each PV module for the system operation. The designed system is used to measure the voltage, current of each PV module, as well as environmental temperature. The master PLC module sends these PV performance parameters to the data logger. The block diagram of the used system is shown in Figure 10.12.

A wireless sensor network system is developed for monitoring the solar PV system performance. Analog sensors such as humidity, temperature, and current sensors are used for single PV modules separately. Performance data is transferred through a radiofrequency (RF) module. Two days performance parameters, such as current of panels 1–3, battery current, and temperature, are transferred from the transmitting side to the receiver side. The layout of the proposed system is shown in Figure 10.13.

In Ref. [17], a grid-connected PV park is considered for real-time monitoring. The obtained results show the utilization of voltage, current, temperature, and irradiation sensors for system protection and extensive monitoring. The performance parameters of solar PV systems, e.g. voltage, current, environmental temperature, and solar irradiation, are measured regularly and monitored at the controlling station/stage using transmitting modules. Simultaneously, the measured performance data are recorded

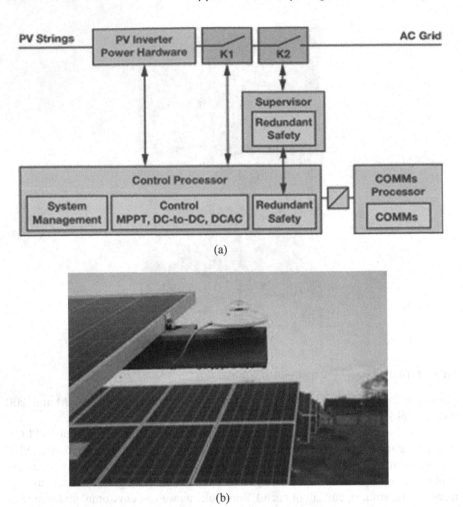

(a)

(b)

FIGURE 10.9 (a) Block diagram of developed data logger. (b) Pyrometer system to measure solar irradiance [12].

for further analysis. In Ref. [18], internet of things (IoT)-based monitoring is carried out to monitor the solar PV performance parameters. The major advantage of the IoT system is that it is easy to monitor a solar plant for performance evaluation remotely.

In Ref. [19], the authors developed an experimental system for monitoring the indoor environmental parameters such as temperature, humidity, human occupancy, light intensity, and air quality monitoring. The measured data are recorded on the memory card (Micro SD card). The prototype model has supportive components such as an Arduino and various analog sensors.

The authors of Ref. [20] developed a low-cost PV analyzer, which comprised the major components of a current sensor (ACS712), voltage sensor (B25), and

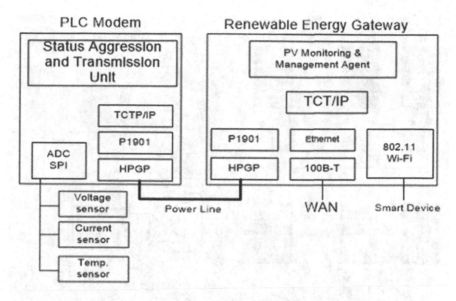

FIGURE 10.10 Structure of PLC modem and renewable energy gateway [13].

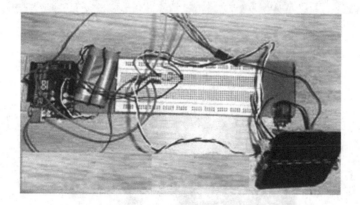

FIGURE 10.11 Block diagram of the developed model of data acquisition system [14].

Atmega328 microcontroller. Moreover, temperature, humidity, and solar irradiance type environment parameters are measured using the analog sensors. The authors considered two separate solar PV modules for online characterization at different irradiation and temperature levels. The experimental setup is shown in Figure 10.14.

A thermal study for the performance improvement of solar PV systems is carried out in this study [15]. An integration of Arduino and supportive sensors–actuator systems is used for the development of the prototype model. An automated water pumping system is driven to control the temperature of the solar PV system, which is validated from the obtained curves of temperature [21]. The developed system is shown in Figure 10.15.

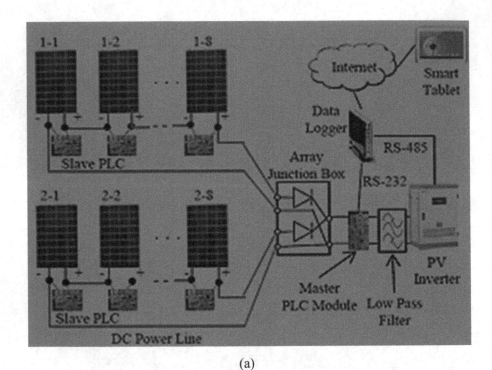

(a)

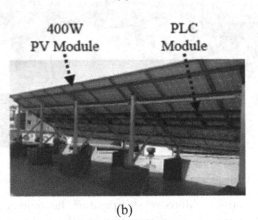

(b)

FIGURE 10.12   (a) Schematic diagram of data logger-assisted PV monitoring system using PLC. (b) Deployed location of PLC [15].

The authors of Ref. [22] measured the temperature for a four-month duration. Two more sensors are connected with the Arduino system for pressure and altitude measurement. The schematic diagram of the developed system is shown in Figure 10.16.

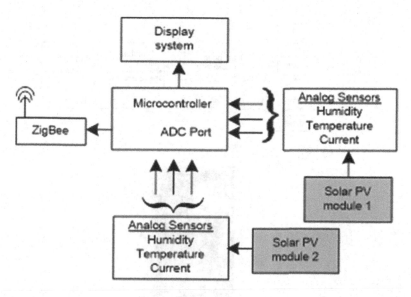

FIGURE 10.13 Layout of PV monitoring system [16].

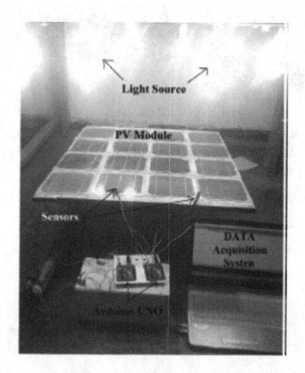

FIGURE 10.14 Experimental setup for online characterization of PV module [20].

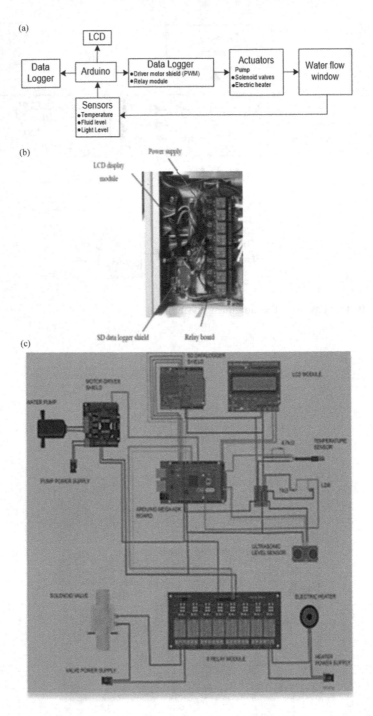

FIGURE 10.15   (a) Block diagram of developed model. (b) Relay circuit control. (c) Schematic diagram of system [21].

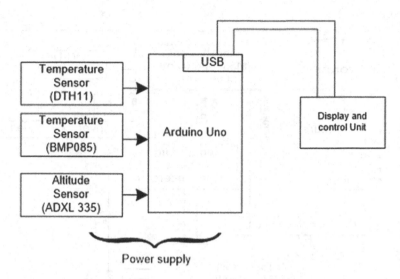

FIGURE 10.16 Schematic diagram of developed system for weather forecasting [22].

In Ref. [23], the authors developed a data logger system to measure the solar PV performance parameters as well as environmental parameters such as solar irradiance, ambient temperature, and humidity. In Refs. [24,25], the authors developed a customized cost-effective system that allows monitoring and predicting the performance of a roof-top installed PV system. Various types of sensors are used for the monitoring of environmental parameters such as ambient temperature, humidity, presence of dust (air pollution monitoring), wind speed, and solar irradiation. For the graphical representation of observed parameters, the Lab-VIEW-based graphical user interface (GUI) system is used and the performance data is transmitted using X-bee and/or Wi-Fi modules. The schematic view of the developed systems are shown in Figures 10.18 and 10.19.

In Ref. [26], the authors have developed a solar harvesting circuit for the wireless HART sensor node. The obtained experimental-based results with the proposed system have shown extensive proof of solar harvesting. In Ref. [27], the authors used analog sensors, such as voltage, current, temperature, and humidity sensors, for the performance investigation of solar PV system. The authors of Refs. [28,29] implemented a real-time performance measurement of a solar PV system. As well as monitoring of performance parameters, data acquisition (DAQ) is interfaced with sensors such as a humidity sensor, temperature sensor, solar irradiation sensor, and voltage and current sensors to store the measured data instantaneously. No data transmitting module is used in the system, and the schematic diagram of the developed system is shown in Figure 10.19.

In view of above literature review, the developed models monitored environmental parameters, e.g. solar irradiation, temperature, humidity, air pollution (dust particles), and performance parameters (voltage and current). Furthermore, it is observed that the performance assessment of the solar PV system is carried

(a)

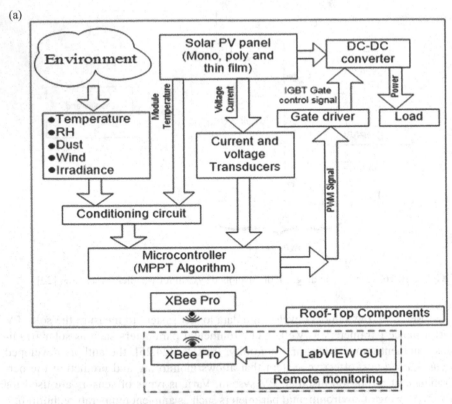

(b)

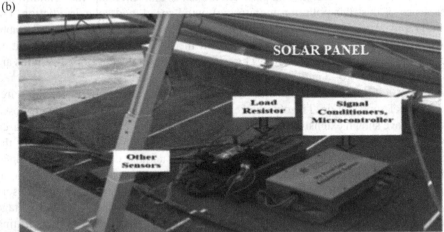

FIGURE 10.17 (a) Schematic diagram of installed PV system monitoring system. (b) Experimental setup comprised of microcontroller circuits, X-Bee Pro, and signal conditioning circuits for the sensors and resistive load [24].

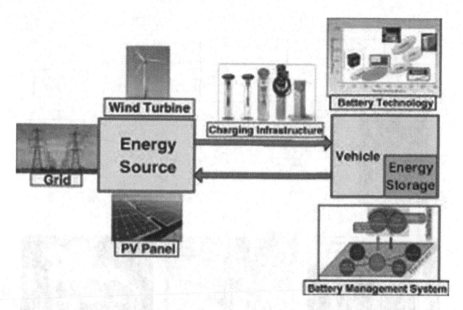

FIGURE 10.18 Layout of developed environmental monitoring system [25].

out accurately and in an automatic manner during the investigation. These models for the solar performance monitoring are analyzed, and all the supportive components used ICT technology, modeling scale, etc. Further details are shown in Table 10.1.

10.3 COMPARATIVE STUDY OF EXISTING SYSTEMS BASED ON TECHNOLOGY AND APPLICATIONS

Table 10.1 shows the technological and experimental/commercial comparison based on applications.

10.4 CONCLUSION

Weather and climatic parameter measurement and monitoring is a challenging domain due to the various difficulties in implementation including technology involved, power management, and deployment considerations. Thus, comprehensive comparative data is of immense utility for the selection of appropriate technology and hardware for deployment in different scenarios. Since in this case, there is no one solution that fits all, the data would assist in assessing the cost-to-benefit ratio as well as the decision-making process. On the hardware side, there is a plethora of choices like Arduino, PIC, etc., with either wired or wireless technology. The status of the technology in terms of being an experimental or a fully commercial solution is also listed, which makes it apt for selection in terms of R&D or physical implementation.

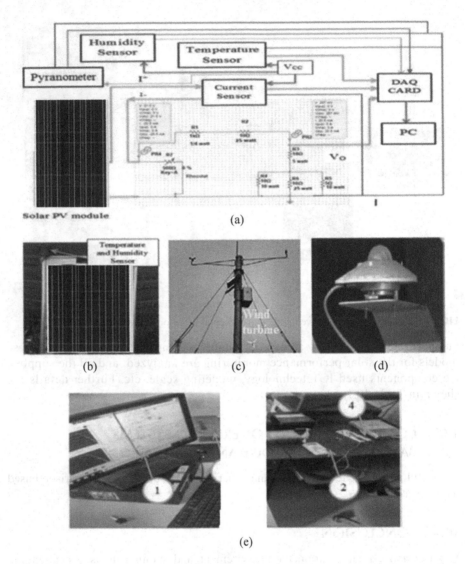

FIGURE 10.19 (a) Schematic diagram of developed system for performance monitoring of PV system. (b) Outdoor PV module. (c) Weather monitoring system. (d) Solar irradiation measurement. (e) Experimental setup and data acquisition system [28].

TABLE 10.1
Comparison of Embedded Based Solar Powered Systems for Environmental Parameters Monitoring

Author's Name, Year, [Refs.]	Major Components of Developed System	Controller/ Technology	Parameters for Measurement	Commercial/ Laboratory- Based Experimental	Wired/Wireless Performance Data Transmission	Application
Aristizabal et al. [1]	• Radiation sensor • Ambient temperature • DC Current sensor • Data acquisition board: PCMCIA 6024E	• SCXI System	• Solar irradiation • Temperature • DC current	Commercial	Wired	Development of monitoring system for performance monitoring and quality of the electrical power generated by the plant.
Forero et al., 2006, [2]	• Temperature sensor • Radiation sensor • Voltage sensor • Current sensor • DAQ board	• PC with Lab-View interface for GUI	• Temperature • Radiation • Voltage • Current	Experimental	Wired	Monitoring of solar PV plant is carried out in terms of performance parameters such as voltage and current. The GUI is shown with Lab-View environment.
Mondal, 2006, [3]	• Data logger • Pyranometers • Anemometer • Temperature sensor	• Controller	• Solar irradiation • Wind speed • Temperature	Commercial	Wired	System performance, maximum power point tracking, automatic switching off–on, and grid monitoring are investigated

(Continued)

TABLE 10.1 (*Continued*)
Comparison of Embedded Based Solar Powered Systems for Environmental Parameters Monitoring

Author's Name, Year, [Refs.]	Major Components of Developed System	Controller/ Technology	Parameters for Measurement	Commercial/ Laboratory-Based Experimental	Wired/Wireless Performance Data Transmission	Application
Drews, et al., 2007, [4]	• Solar irradiation sensor • Temperature sensor	NA	• Solar irradiation • Temperature	Commercial	Wired and satellite-based monitoring	A small grid-connected PV system is developed and smart performance assessment is carried out.
Naeem, et al., 2011, [5]	• Temperature sensor • Voltage sensor • Local area network	• Microcontroller: PIC18	• Temperature • Open circuit voltage	Commercial	Web server LAN	The architecture and construction of a simple embedded web server system and its application for monitoring of a PV system
Rashidi, et al., 2011, [6]	• Voltage sensor • Current sensor	• Microcontroller • LAB-View: Graphical user interface (GUI)	• Voltage • Current • Power	Commercial	Zigbee	Wireless monitoring-based system is developed for monitoring of the PV array system.
Krejcar and Mahdal, 2012, [7]	• Voltage, sensor • Current sensor • Buck voltage converter • Data logger	NA	Power	Experimental	Wireless module JN5139	A wireless-based data logger is designed to assess the performance of the designed system

(*Continued*)

TABLE 10.1 (*Continued*)
Comparison of Embedded Based Solar Powered Systems for Environmental Parameters Monitoring

Author's Name, Year, [Refs.]	Major Components of Developed System	Controller/ Technology	Parameters for Measurement	Commercial/ Laboratory-Based Experimental	Wired/Wireless Performance Data Transmission	Application
Lopez, et al., 2012, [8]	• Solar irradiation sensor • Wind speed sensor • Voltage and current sensors • UART: RS-232	• Microcontroller unit (MCU): dsPIC 30F3013	• Solar irradiation • Wind speed • Voltage and current of PVA	Commercial	RF transceiver	Metrological parameters and solar PV performance measurement in terms of voltage and current system are investigated.
Arkoub, et al., 2013, [9]	• Voltmeter • Ammeter • Temperature sensor • Irradiation sensor • Data logger	• RS-232 interface to a PC • Lab-VIEW for GUI	• Voltage • Current • Temperature • Irradiation	Experimental	Wired (RS-232 interfacing)	A wired experimental system is designed to sense the voltage, current, temperature, and solar irradiation of the PV system performance evaluation. Moreover, a Lab-VIEW software-based tool box is used for the GUI of proposed system.

(*Continued*)

TABLE 10.1 (*Continued*)

Comparison of Embedded Based Solar Powered Systems for Environmental Parameters Monitoring

Author's Name, Year, [Refs.]	Major Components of Developed System	Controller/ Technology	Parameters for Measurement	Commercial/ Laboratory-Based Experimental	Wired/Wireless Performance Data Transmission	Application
Papageorgas, 2013, [10]	• Voltage and current sensor • Temperature sensor • Humidity sensor: SHT11	• Arduino board	• Voltage and current • Temperature • Humidity	Laboratory scale	Zigbee	A laboratory-based hardware system is developed to monitor the short-circuit and open-circuit voltages. Moreover, temperature and humidity are also monitored.
Dyreson, et al., 2014, [11]	• Irradiation sensor (45 no.): LI-COR • Data logger: 2 GB	Microcontroller	Solar irradiation	Commercial	Wired	The irregular behavior of solar irradiation causes the performance reduction of solar PV system; thus, 45 irradiation sensors are located to identify the variation in solar irradiation.

(Continued)

TABLE 10.1 (Continued)

Comparison of Embedded Based Solar Powered Systems for Environmental Parameters Monitoring

Author's Name, Year, [Refs.]	Major Components of Developed System	Controller/ Technology	Parameters for Measurement	Commercial/ Laboratory-Based Experimental	Wired/Wireless Performance Data Transmission	Application
Fuentes, et al., 2014, [12]	• Temperature sensor • Irradiance sensor • Data logger: Agilent34970A • Real time clock	Arduino	• Temperature • Solar irradiance	Experimental	Wired	A low-cost solar PV performance monitoring system is designed and a data logger assists the proposed system to store the measured data in terms of voltage, current, etc.
Han, et al., 2014, [13]	• Voltage, current sensor • Temperature sensor	• Microcontroller unit	• PV performance monitoring	Commercial	PLC modems, WAN, Wi-Fi	A smart device application is developed and used to display the performance status of the entire PV system.
Gad and Gad, 2015, [14]	• Temperature sensor (16 no.) • SD card • Liquid crystal display (LCD) • Real Time Clock (RTC)	• Arduino: ATMega 2560, • Serial peripheral interface: PIC18F46K20	• Ambient Temperature of solar system	Commercial	Wired	DAS for temperature monitoring in solar installations

(Continued)

TABLE 10.1 (Continued)
Comparison of Embedded Based Solar Powered Systems for Environmental Parameters Monitoring

Author's Name, Year, [Refs.]	Major Components of Developed System	Controller/ Technology	Parameters for Measurement	Commercial/ Laboratory-Based Experimental	Wired/Wireless Performance Data Transmission	Application
Han, et al., 2015, [15]	• Temperature sensor • Voltage sensor • Current sensor • Data logger	• Microcontroller unit	• Temperature • Voltage • Current	Commercial	• Power line communication (PLC) • Wi-Fi	A lost-cost and user-friendly system to monitor the performance of a PV system is designed and deployed.
Kaundal, et al., 2015, [35]	• Current sensor: ACS712 • Temperature sensor • Humidity sensor • LCD	Microcontroller	Voltage	Experimental	Zigbee RF module: CC2500	Performance decrement of the solar PV system is monitored by a wireless sensor node (Zigbee-RF module)
Moreno-Garcia, et al., 2015, [17]	• Temperature sensor: PT100 • Voltage sensor: LV 25-P • Current sensor: LEM LA 305-S • Irradiation sensor	• Real time programmable controller	• Temperature • Voltage • Current • Irradiation	Commercial	• Radio frequency • GPS	The proposed system monitors several parameters and ensures the performance quality based on retrieval and storage.
Adhya, et al., 2016, [18]	• Temperature sensor: LM35 • Voltage and current transducers	Microcontroller: PIC18F46K22	Power	Experimental	GPRS module: SIM900A	An IoT-based experimental system is designed to monitor the solar PV system.

(Continued)

TABLE 10.1 (Continued)
Comparison of Embedded Based Solar Powered Systems for Environmental Parameters Monitoring

Author's Name, Year, [Refs.]	Major Components of Developed System	Controller/ Technology	Parameters for Measurement	Commercial/ Laboratory-Based Experimental	Wired/Wireless Performance Data Transmission	Application
Ali, et al., 2016, [19]	• micro SD card • RTC: DS3234 • Light intensity: TSL2561 • Proximity sensor • CO_2 concentration	• Arduino: Atmel Atmega 328P	• Temperature • Humidity • Human occupancy • Light intensity • CO_2 concentration	Prototype	Wired	Arduino platform for measuring and recording long-term indoor environmental and building operational data
Anand, et al., 2016, [20]	• Irradiance sensor • Temperature and humidity sensor • Current sensor • Voltage sensor • Resistive load	Arduino	Voltage and current	Experimental	Wired	A low-cost Arduino-based system with the assistance of sensors is designed to measure the current and voltage of solar PV module.
Claros-Marfil, et al., 2016, [21]	• Temperature sensor: DS18B20 • SD card • Ultrasonic sensor (SR04) • 8 relay circuits: 30V DC operated • LCD: 16x2 • Light dependent resistor • Water pump • Fluid level sensor	• Arduino: Atmel Atmega ADK	• Water flow • Temperature	Prototype	Wired	Temperature control of tank during water flow

(Continued)

TABLE 10.1 (Continued)
Comparison of Embedded Based Solar Powered Systems for Environmental Parameters Monitoring

Author's Name, Year, [Refs.]	Major Components of Developed System	Controller/ Technology	Parameters for Measurement	Commercial/ Laboratory-Based Experimental	Wired/Wireless Performance Data Transmission	Application
Laskar, et al., 2016, [22]	• Temperature sensor: DHT-11 • Pressure sensor: BMP-085 • Accelerometer: ADXL-335	Arduino Uno Atmel Atmega 328 microcontroller	• Pressure • Humidity • Temperature	Prototype	RF module-433 MHz	Environmental parameters, e.g. pressure, temperature, and humidity, are measured at different altitudes
Shukla, et al., 2016, [23]	• Ammeter and voltmeter • Infrared gun • Thermometer • Irradiance meter: TM 207 • Environmental meter • Variable resistive load	Voltage and current measurement using Multimeter: RM 155	• Energy and exergy efficiency • Temperature • Humidity	Experimental	Wired	A transparent PV module is considered for the performance assessment under humidity and separate temperature environment
Touati, et al., 2016, [24]	• RTD sensor • Wind speed sensor • Dust sensor	Microcontroller	• Ambient temperature • Humidity • Dust • Wind • Solar irradiance	Prototype	Xbee Pro	Assessment of environmental parameters, e.g. ambient temperature, dust accumulation, humidity, solar irradiation, etc., and the effect on PV performance

(Continued)

TABLE 10.1 (Continued)
Comparison of Embedded Based Solar Powered Systems for Environmental Parameters Monitoring

Author's Name, Year, [Refs.]	Major Components of Developed System	Controller/ Technology	Parameters for Measurement	Commercial/ Laboratory-Based Experimental	Wired/Wireless Performance Data Transmission	Application
Wong, et al., 2016, [25]	• Humidity sensor • Temperature sensor • Dust intensity sensor • Ultra Violet • Radiation sensor • Noise index sensor • LCD display	Arduino Uno Atmel Atmega	• Humidity • Temperature • Dust intensity • Ultra Violet radiation • Noise index	Prototype	GPS, Wi-Fi and 3G	Development and deployment of microenvironmental monitoring system at construction sites
Ibrahim, et al., 2016, [26]	• Voltage regulator: LM317 • Heat sensor	Arduino Mega	Voltage	Experimental	Wireless sensor node	Wireless sensor node-based solar performance data transfer system is designed.
Patil, et al., 2017, [27]	• Current sensor: ACS712 • Voltage sensor • Resistors (10–100 kΩ)	Arduino	• Voltage • Current	Experimental	IoT	The sensor-based circuit system is designed with Arduino and used for communicating the performance of solar PV system on the cloud through IoT.

(Continued)

TABLE 10.1 (*Continued*)
Comparison of Embedded Based Solar Powered Systems for Environmental Parameters Monitoring

Author's Name, Year, [Refs.]	Major Components of Developed System	Controller/ Technology	Parameters for Measurement	Commercial/ Laboratory-Based Experimental	Wired/Wireless Performance Data Transmission	Application
Rohit, et al., 2017, [28]	• Humidity sensor • Temperature sensor • Irradiation sensor • Current sensor • Voltage sensor • DAQ card	Arduino PC interface, Lab-VIEW	• Humidity sensor • Temperature • Irradiation • Current • Voltage	Experimental	Wired	Real-time data acquisition and monitoring of 5 W PV module is carried out in the energy center. Different parameters are successfully acquired and displayed on the front panel of the LabVIEW software.
Song and Haberl, 2017, [29]	• Pyranometer • Data logger	PC interface	• Solar irradiation • Solar transmittance	commercial	Wired	Measurement and validation on-site global solar transmittance as a function of varying angles of incidence for glazing samples under natural clear-sky conditions is carried out.

REFERENCES

1. Aristizabal, A. J., Arredondo, C.A., Hernandez, J., Gordillo, G., 2006 Development of equipment for monitoring PV power plants, using virtual instrumentation. *IEEE 4th World Conference on Photovoltaic Energy Conference.*
2. Forero, N., Hernandez, J., Gordillo, G., 2006 Development of a monitoring system for a PV solar plant. *Energy Conversion and Management* 47(15–16); 2329–2336.
3. Mondol, A. J. D., Yohanis, Y., Smyth, M., Norton, B., 2006. Long term performance analysis of a grid connected photovoltaic system in Northern Ireland. *Energy Conversion and Management* 47; 2925–2947.
4. Drews, A. C. D., Keizer, H. G., Lorenz, B. E., Betcke, J., Sark, W. G. J. H. M. V., Heydenreich, W., Wiemken, E., Stettler, S., Toggweiler, P., Bofinger, S., Schneider, M., Heilscher, G., Heinemann, D., 2007. Monitoring and remote failure detection of grid-connected PV systems based on satellite observations. *Solar Energy* 81; 548–564.
5. Naeem, M., Anani, N., Ponciano, J., Shahid, M., 2011 Remote condition monitoring of a PV system using an embedded web server. *IEEE PES Innovative Smart Grid Technologies Conference Europe (ISGT Europe).*
6. Rashidi, Y., Moallem, M., Vojdani, S., 2011 Wireless Zigbee system for performance monitoring of photovoltaic panels. *IEEE Conference on Photovoltaic Specialists.*
7. Krejcar, O., Mahdal, M., 2012. Optimized solar energy power supply for remote wireless sensors based on IEEE 802.15.4 Standard. *International Journal of Photo Energy.*
8. Lopez, M. E. A., Mantiian, F. J. G., Molina, M. G., 2012 Implementation of wireless remote monitoring and control of solar photovoltaic (PV) System.
9. Arkoub, V., Alkama, R., Blum, K., 2013. Measurement array for photovoltaic installation. *Energy Procedia* 36; 211–218.
10. Papageorgas, P., Piromalis, D., Antonakoglou, K., Vokas, G., Tseles, D., Arvanitis, K. G., 2013. Smart solar panels: In-situ monitoring of photovoltaic panels based on wired and wireless sensor networks. *Energy Procedia* 36; 535–545.
11. Dyreson A. R., Morgan, E. R, Monger, S. H., Acker, T. L., 2014. Modeling solar irradiance smoothing for large PV power plants using a 45 sensor network and the wavelet variability model. *Solar Energy* 110; 482–495.
12. Fuentes, M., Vivar, M., Burgos, J. M., Aguilera, J., Vacas, J. A., 2014. Design of an accurate, low-cost autonomous data logger for PV system monitoring using Arduino that complies with IEC standards. *Solar Energy Materials and Solar Cells* 130; 529–543.
13. Han, J., Lee, I., Kim, S. H., 2015. User friendly monitoring system for residential PV system based on low cost power line communication. *IEEE International Conference on Consumer Electronics (ICCE).*
14. Gad, H. E., Gad, H. E., 2015. Development of a new temperature data acquisition system for solar energy applications. *Renewable Energy* 74; 337–343.
15. Han, J., Choi, C. S., Park, W. K., Lee, I., Kim, S. H., 2014. PLC based photovoltaic system management for smart home energy management system. *IEEE International Conference on Consumer Electronics.*
16. Kaundal, V., Mondal, A. K., Sharma, P., Bansal, K., 2015. Tracing of shading effect on underachieving SPV cell of an SPV grid using wireless sensor network. *Engineering Science and Technology, an International Journal* 18; 475–484.
17. Moreno, G. I. M., Pallares, L. V., Gonzalez, R. M., Lopez, L. J., Varo, M. M., Santiago I., 2015. Implementation of a real time monitoring system for a grid connected PV Park. *IEEE Conference, 2015.*
18. Adhya, S., Saha, D., Das, A., Jana, J., Saha, H., 2016. An IoT based smart solar photovoltaic remote monitoring and control unit. *International Conference on Control, Instrumentation, Energy & Communication.* doi:10.1109/CIEC.2016.7513793.

19. Ali, A. S., Zanzinger, Z., Debose, D., Stephens, B., 2016. Open source building science sensors: A low-cost arduino-based platform for long term indoor environmental data collection. *Building and Environment* 100; 114–126.
20. Anand, R., Pachauri, R., Gupta, A., Chauhan, Y. K., 2016. Design and analysis of a low cost PV analyzer using Arduino UNO. *IEEE Conference*.
21. Claros-Marfil, L. J., Padial, J. F., Lauret, B., 2016. A new and inexpensive open source data acquisition and controller for solar research: Application to a water-flow glazing. *Renewable Energy* 92; 450–461.
22. Laskar, M. R., Bhattacharjee, R., Giri, M. S., Bhattacharya, P., 2016. Weather forecasting using Arduino based cube-sat. *Procedia Computer Science* 89; 320–323.
23. Shukla, A. K., Sudhakar, K., Baredar, P., 2016. Exergetic analysis of building integrated semi transparent photovoltaic module in clear sky condition at Bhopal India. *Case Studies in Thermal Engineering* 8; 142–151.
24. Touati, F., Al-Hitmi, M.A., Chowdhury, N. A., Hamad, J. A., Antonio J.R., Gonzales, S. P., 2016. Investigation of solar PV performance under Doha weather using a customized measurement and monitoring system. *Renewable Energy* 89; 564–577.
25. Wong, M. S., Mok, E., Wang, T., Yong, Z., 2016. Development of an integrated micro-environmental monitoring system for construction sites. *Procedia Environmental Sciences* 36; 207–214.
26. Ibrahim, R., Chung, T. D., Hassan, S. M., Bingi, K., Salahuddin, S. K. B., 2017. Solar energy harvester for industrial wireless sensor nodes. *Procedia Computer Science* 105; 111–118.
27. Suprita, M. Patil, Vijayalashmi, M., Tapaskar, R., 2017. Solar energy monitoring system using IoT. *Indian Journal Science Research* 15(2); 149–155.
28. Rohit, A. K., Tomar, A., Kumar, A., Sangnekar, S., 2017. Virtual lab based real-time data acquisition, measurement and monitoring platform for solar photovoltaic module. *Resource-Efficient Technologies* 3(4); 1–6.
29. Song, S., Haberl, J. S., 2017. Simplified field measurement and verification of global solar transmittance for glazing samples under natural clear sky conditions. *Solar Energy* 155; 706–714.

11 IoT with Automation Clustering to Detect Power Losses with Energy Consumption and Survey of Defense Machinery against Attacks

Arun Kumar Singh
Saudi Electronic University, Saudi Arabia-KSA

Vikas Pandey
Babu Banarasi Das University, Lucknow, India

CONTENTS

11.1 INTRODUCTION

Losses or loss of electrical power in the system electric power distribution, which is usually used at certain times, is one measure that is efficient or not an electric power system operation. Currently, it has been applied the method of measuring electrical energy using the AMR (Automatic Meter Reading) system, namely, a system of reading or retrieving data from the measurement of electrical energy on consumers locally and remotely, where the reading schedule can be determined as needed. This AMR system can be used optimally for account issuance, customer load analysis, calculation of losses or distribution losses, and electricity network development planning [1].

The data will be processed and grouped based on the usage pattern of its power. This study uses the K-Means Clustering Data Mining method. The K-Means method is the oldest and most widely used clustering algorithm in a variety of small to medium applications because of its ease of implementation and fast time complexity. Besides that, K-Means also has a fairly high accuracy on the size of objects, so this algorithm is relatively more measurable and efficient for processing large amounts of objects. The weaknesses in the K-Means algorithm are analyzing and determining the best number of k in clustering data in a dataset. To get the optimal k value, the author uses the Davies–Bouldin Index (DBI) method. The DB Index measures the average similarity between each cluster and one of the most similar.

Internet of Things (IoT) is technology where devices get a particular identity to be able to connect and communicate with one another through internet networks without human-to-human or human-to-machine interactions. A collection of IoT devices connected by a network is called IoT infrastructure. It is group into four layers, i.e., sensors and actuators, internet gateway and data acquisition system, edge handler, and data center. Sensors and actuators are used to collect data from the environment or physically observed objects. Units of sensors and actuator are what we call nodes. Each node can communicate with the other using specific protocols to produce this useful set of data. Analog data from sensors and actuators are converted into a digital form by data acquisition devices, which is then forwarded by the internet gateway to Edge handler layer. The edge handler's function is to prevent data from the edge from consuming the data center bandwidth. It can also process raw data into data that is ready to be processed. The last layer is Data Center and Cloud; at this layer, data is processed and analyzed in-depth for later use by its users [2].

In contrast to the current paradigm on the Internet, which bases on human-to-human relations, Gutiérrez mentioned IoT as having a paradigm as the future internet, where every physical or virtual object that can be identified with unique identifiers will be considered to be interconnected. So, keeping this in mind, although IoT uses distributed networks in nature, IoT has driven combinations with other technologies, such as short-range communication, real-time localization, embedded sensors, and ad-hoc networks as a way to turn everyday things into smart things. Combining IoT with an ad-hoc network provides benefits because of the ad-hoc properties; as self-organized networks, they are built spontaneously by several connected devices. on a router or base station, so they are suitable for implementation where the deployment of new fixed infrastructure is not feasible.

In addition, when the mobility characteristic is calculated, it becomes a wireless ad-hoc network. Wireless ad-hoc network itself represents a new communication paradigm where decentralized wireless nodes communicate with each other in collaborative ways to achieve common goals. So, considering the many capabilities of the wireless ad-hoc network, it would be highly beneficial to IoT, and this will also be suitable for implementations that require mobility [3].

On the other hand, there is something to be considered in the integration of these two technologies. As both depend on the node that communicates using particular identities, both are still facing common security problems. It is vulnerable to Sybil attack, defined as an intrusion where malicious devices get or change into several different identities illegally. Based on its characteristics, Sybil attack is also grouped into the identity-based attack, which attacks compromised systems using false identities. This type of attack disguises itself as legitimate devices, and it is done by attacker camouflage its intrusion packet data similar to regular data packets. The security system would find it difficult to distinguish between the two types of data packages. For detecting this kind of attack, a lot of traditional countermeasures are proposed. However, adopting traditional security countermeasure cannot effectively be used in IoT due to its source limitation. Along with the many studies regarding the method of securing Sybil attacks on the wireless ad-hoc network, the question that arises is related to what methods are used in the IoT defense mechanism and what is the drawback. Similar questions have been investigated by focusing on non-IoT infrastructure. In this paper, researches related to machine learning on IoT security are collected from various sources and then reviewed using the systematic literature review method.

its integration in IoT infrastructure that has limitation in the resource. So the aim of this paper is to present a survey of security mechanism that has been proposed for wireless ad-hoc network to get the results of the analysis of which methods are suitable and what needs to be taken into account in the implementation of security machinery in IoT; we classify each paper then analyze advantages and limitations to assess which methods are suitable to be implemented in the wireless ad-hoc networks application in IoT [4].

11.2   RESEARCH METHOD

The data used in this study is the historical data of customer power usage of AMR (Automatic Meter Reading), electricity tariffs for business purposes, 'B-3 tariff rates, or medium voltage (power limits above 200 kVA). The attributes that will be used are Power based on Peak Load Time and Power based on Outdoor Peak Load Time. Data training is used by customers with normal power electricity usage. Data testing is used by 3 customers who are classified as having non-normal power electricity usage (non-technical losses).

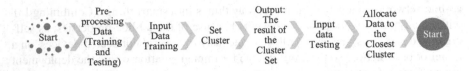

FIGURE 11.1 Design process.

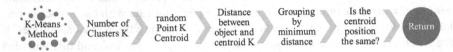

FIGURE 11.2 K-Means process.

The research methods are shown in Figure 11.1.

This stage is the majority work in data mining applications. This stage is the process of understanding data and initial data processing, the results of which will later be used in the calculation process with predetermined methods. The data to be processed is a load profile data on the power usage of each AMR customer for business customers of the B3/TM tariff group [5]. Next is the calculation using the K-Means method to determine the group pattern of AMR (Automatic Meter Reading) customer power usage business class [6]. The calculation stages will be explained based on the flowchart process shown in Figure 11.2.

The following is an example of the calculation process for centroid grouping, going directly to the last iteration stage; the data can be seen in Table 11.1.

11.2.1 SYBIL ATTACK PROPERTIES

Sybil attack is defined as an intrusion where malicious devices get or change into several different identities illegally. Newsome convey the impact of the Sybil attack on several protocols, including: Distributed storage: when there are nodes that cannot provide services, then the node will share its data to neighboring nodes. If this neighbor node is a Sybil node, then the data can be obtained. Routing: especially a network that has a sink, when the Sybil point has gained control of the sink node; in

TABLE 11.1
Centroid Set 2 Cluster Final

	A	B	C	D	E	F	G
C1	0.266	0.250	0.284	0.252	0.279	0.247	0.273
C2	0.630	0.476	1.213	1.220	1.244	1.250	1.151

	H	I	J	K	L	M	N
C1	0.247	0.276	0.246	0.267	0.247	0.266	0.255
C2	1.249	1.227	1.231	1.197	1.304	0.635	0.808

addition to the Sybil node getting all data passed on the network, many other attacks can be carried out [7]. Data aggregation: if the Sybil node mediates data packets, then it can manipulate the data. Voting: by increasing the number of nodes, Sybil nodes can influence the results of the voting, or as most Sybil points can accuse legitimate points of being evil. Fair allocation of resources: the Sybil node can disrupt the system by unauthorized activation/deactivation of the node. So as to avoid these impacts, defense machinery that can accurately detect Sybil attacks are needed. Mishra classifies Sybil attacks.

11.2.2 COMPROMISED PHASE

The attacker tries to get a group of nodes that can be controlled by the attacker. There are two characteristics of Sybil attack at this stage, according to the way the attacker gets a node to be able to enter the network, namely by making stolen/compromise and by doing fabrication. This phase ends when the attacker gets a group of compromised nodes that are connected in the destination network [8].

- Fabrication: characteristics of Sybil attacks with Fabrication are usually carried out when there is the possibility of the attacker to create a new identity in accordance with network requirements. For example, if the network only gives an ID in the form of a number of n-bits, the Sybil attacker can create a new random identity randomly within a valid range $(0–n)$ so that it is recognized as a valid node.
- Stolen/Compromised: if a fabrication attack cannot be carried out, then what the attacker can do is to steal the identity of a valid node. If one of the nodes or a group of valid nodes in the network can be taken over, the attacker can use this node directly, or by taking its identity then the attacker temporarily interferes with the valid node or destroys it permanently.

11.2.3 DEPLOYMENT PHASE

Sybil attackers will try to spread the nodes that are taken over by gathering network-related information. The most crucial thing in this phase is that the attacker will determine the placement of compromised nodes in strategic locations and allow for success in the launch phase. Sybil nodes can be moved at specific locations to be able to attack simultaneously, or individual nodes can be endeavored to take on the role of cluster heads. There are two characteristics of Sybil at this stage, according to the capabilities of the Sybil attacker, namely spreading randomly and selectively [9].

- Random Deployment: the attacker chooses a location to use Sybil randomly.
- Selective Deployment: the attacker selectively chooses the set of Sybil nodes it has. For example, deploying the group at one central location so that it can dominate that location, or the attacker can spread Sybil nodes to various places on the network to avoid behavior-based detection.

11.2.4 LAUNCHING PHASE

There are several forms of Sybil attacks in carrying out attacks. This is adjusted to the objectives to be achieved by the attack, whether to disrupt the system, do the DoS, or other objectives. Forms of attack that are launched are also usually intended to avoid detection systems. The attack can be carried out directly, i.e., the Sybil node communicates directly with the valid node, or indirectly, i.e., the attack is carried out by communication through one of the Sybil nodes. Indirect Communication: in this attack version, there is no node that can communicate directly with Sybil. Instead, one or more malicious devices are claimed to have reached Sybil's point. Messages sent to the Sybil node are routed through one of these dangerous intersections, which pretends the message is returned to the Sybil node [10].

- Simultaneous: Attackers deploy a group or all Sybil nodes simultaneously. This group can directly connect with the network or participate through other Sybil points.
- Non-simultaneous: Attackers do not attack simultaneously; for example, the attacker can choose attacks alternately according to a specific time lag. Usually, this is done to avoid specific detection.
- Conspiracy Sybil: Sybil attacks that conspire to do by attacking the Sybil node network will freely control nodes that are compromised by other points as accomplices to attack directly or by using these nodes to give new identities to other Sybil nodes. The Sybil conspiracy attack in vehicle ad-hoc network (VANET) was first introduced wherein the attacker could pretend to be a conspiracy node and then use this identity to send malicious messages to other nodes nearby

The following is an example of the calculation process for centroid grouping, going directly to the last iteration stage; the data can be seen in Table 11.1.

After doing K-Means calculations and getting a centroid grouping, the following is an example of the data to be tested. The developed system is shown in Tables 11.2–11.4.

Calculate the maximum distance value (max DC) in each cluster that is formed. (For example, the optimal cluster set = 2, based on the previous stage.)

Calculate the DC of each training data object to centroid, using the Euclidean formula.

TABLE 11.2
Examples of Data Testing

Customer	A	B	C	D	E	F	G
1	0.374	0.342	0.398	0.347	0.402	0.347	0.391
Customer	H	I	J	K	L	M	N
1	0.344	0.400	0.352	0.390	0.352	0.376	0.363

TABLE 11.3

Maximum DC in Each Cluster

Customer	DC1	DC2
1	0.418	
2	0.148	
3	0.292	
4		0
5	0.273	
MAX DC	0.418	0

TABLE 11.4

DC Value

DC1	DC2
0.409	2.776

Compare the value of DC data testing with each max DC in each cluster that has been calculated. The results of the appeal can be categorized as follows:

1. If all DC data testing ≤ max (DC) per cluster, then the data is classified as normal in use.
2. If one of the DC data testing ≥ max (DC) per cluster, then the data is classified as normal in use.
3. If all DC data testing ≥ max (DC) per cluster, then the data is classified as abnormal data in use.

DC1 Testing: DC1 Training = 0.409 < 0.418.
DC2 Testing: DC2 Training = 2.776 > 0.

Based on the results of the above comparison, it can be concluded, that Data Testing is classified as **Normal Data** in power usage. From the calculation process described above, the sample used is 5 AMR Business Class customers (B-3) [11] by grouping using cluster set 2 and using 14 attributes in which 14 attributes are divided into 2, 7 of which are LWBP power usage data for 7 7 days and again is the data of WBP power usage for 7 days, in its calculations get results, namely:

1. There are 2 patterns formed from the results of 5 customer data clustering. The power usage pattern is formed from the final centroid value obtained after the result of the iteration ends.
2. In the first cluster, there are 4 customers, while for the second cluster, there is only 1 customer. That is, of the 5 customers, there are 4 customers who have similarities in the pattern of power usage and 1 other customer has a different pattern from the 4 customers.

3. For the optimal level of the set of 2 clusters for the 5 customers based on the calculation, we obtained a value of 0.095 based on calculations using the DB Index formula.

After the pattern is obtained, a testing test is performed to later make a comparison on whether the testing data is entered or shaped like the pattern obtained or not. If the results of the testing data do not have a similarity with the pattern, in this case the distance to the centroid of each cluster that is obtained exceeds the maximum distance of the members in the cluster, then the customer testing data is unfair in the use of electric power and is set as the operating target [12].

11.3   ATTACK DETECTION IN WIRELESS AD-HOC NETWORK

The defense mechanism of Sybil by considering the characteristics of Sybil that has been mentioned is vital to improve detection accuracy. We have reviewed several defense machineries from Sybil attacks on wireless ad-hoc networks using the SLR method. In general, the steps taken are planning, implementing, and documenting. The planning step consists of identifying review needs, defining and taking specific research questions, developing research protocols, and evaluating review protocols. In the second stage, the implementation of research identification is carried out by conducting a pilot selection and extraction, followed by a selection of the main study quality assessment, data extraction, and data synthesis. The last step taken is documentation, including drawing conclusions and considering threats [13].

11.3.1   CRYPTOGRAPHIC BASED

This method uses the cryptographic protocol, often mainly used to prevent the occurrence of Sybil attacks. Broadly speaking, this defense mechanism is performed by authenticating nodes, using public key certificates to guarantee trust, using secret symmetric keys to prevent other nodes from communicating with the network, and using watermarking to guarantee valid data.

- Authentication: the schema working with each node must be able to prove that it is a valid node through a series of message exchanges on the authentication protocol.
- Public Key Infrastructure: a cryptographic system based on public keys is used to improve security by allowing nodes to communicate in networks with trust values based on certificates held. In this system, certificate-based techniques are used in encryption and authentication machinery. Centralized authority for certification is required.
- Symmetric Key: this scheme relies on encrypting and decrypting messages between nodes using a symmetric encryption algorithm. This technique is used in the network to create secure paths to communicate with each other by using a set of pre-agreed keys or using a trusted third party to ensure the distribution of keys to all legitimate nodes in the network. With this defense

mechanism, the Sybil node will have difficulty getting the key so that it is only possible to obtain a compromised node by stealing [14].

- Watermarking: Watermarking techniques used to be the solution to implementing cryptography on devices with limited resources. The main idea is to embed information that allows an individual to add verification messages to communication data. So, the Sybil nodes cannot make an attack because it cannot change the watermark constraints that have been embed in data.

In the application of IoT defense machinery using cryptography, there are disadvantages:

- Dependence on cryptographic hardware and software.
- Compatibility issue with network types and routing protocols on IoT.
- Scalability in the addition of new nodes/points that may increase resource requirements exponentially.
- High memory, computing, and communication overhead that is not suitable for resource-constrained network.
- To ensure the network has safe keys and algorithms, high costs are needed for key generation and key distribution.

11.3.2 LOCATION VERIFICATION BASED

- The location/position-based method utilizes measurement parameters that can be physically observed to estimate the location and position of the node to detect Sybil attacks. This method is used with the assumption that there may not be different nodes that are in the same location. So that if found, it will be concluded as a Sybil node. Another assumption is to use position verification where a node equipped with a Global Positioning System (GPS) will send its location to a valid node, and then the node will verify based on the estimated position of the propagation model of the received signal [15].
- This method can be grouped into two categories, namely, range-based and range-free methods.
- Range-based: the estimated position is calculated based on the physical indicator used to estimate the distance between the transmitter and receiver. This distance estimate is usually based on the Received Signal Strength Indicator (RSSI), time-based methods such as Time of Arrival (ToA) and Time Difference of Arrival (TDoA). This method is suitable for IoT devices because it is low cost, where the distance between two entities is estimated only based on the received signal strength and the indicators that the device has by default.
- Range-free: this method has high accuracy in distance calculation. By utilizing data from GPS, Radar, or location-based/localization scheme, this method can also be used as a support for position estimation using ranged based.

In applying IoT, the location/position-based defense method has disadvantages. Location/Position-based defense method is not suitable for use on mobile networks such as MANET and VANET; the accuracy of approximate location decreases due to rapid changes in network topology and changes in node position.

Location/Position-based accuracy of the method depends on the environment. Interference, multipath fading, and shadowing lead to inaccurate location estimation.

Location/Position-based privacy violations where identity is required to send position information so that the route of movement of the nodes can be traced.

11.3.3 NETWORK BEHAVIOR-BASED

This method purely detects Sybil nodes based on their features and behavior in the network. The detection method specifically detects features that allow accurate classification between Sybil nodes and valid nodes [16].

In applying IoT, network behavior-based defense method has disadvantages including:

- Only detects Sybil nodes according to the context expected by the detection method, so that Sybil nodes with specific knowledge can escape detection.
- It requires specialized hardware that has a large capacity to collect and analyze data.

11.3.4 RESOURCE TESTING

This method approach is made by testing the unique resources of the node, assuming that each physical node has specific limited resources. A node will be challenged to provide knowledge about specific resources (usually in the form of physical fingerprinting or based on energy), then the verifier compares the resources used by an entity with the typical value or threshold of the resources owned by that entity. Incompatibility indicates the possibility of a Sybil attack [17–19].

- Energy-based: the basic idea of energy-based testing is to verify assuming the node has a predictable energy parameter, so that if a node is found to be incompatible with the existing node in providing an answer, then the node is considered a malicious node.
- Physical fingerprinting: each device has unique characteristics. This characteristic is the basis of verification to determine whether the point is valid or not.

11.3.5 TRUST-BASED

Trust is defined as a relationship of trustor and trustee; the trustor can periodically evaluate the trusteeship to assess its eligibility. Trusted based is based on the value of trust that must be maintained by each node to remain in the network. This trust value can be obtained from trusted devices or from neighbor trusts [20].

- Centralized trust, In the trust-based method using a trusted device, usually in the initial stage, a comprehensive network mapping is carried out on all nodes, with the device obtaining its identity and trust value. Then, the trust value is evaluated to determine the possibility that the node is not a Sybil node.

• Decentralized trust, In the detection approach based on the relationship between neighbors, each node will visit nearby nodes based on the pattern of relationships and behavior of these nodes in the network.

In applying IoT, the trust-based defense method has disadvantages, including the method is not able to detect if Sybil node dominates the number of nodes in the process of determining the value of trust.

The defense mechanism of Sybil by considering the characteristics of Sybil that has been mentioned is essential to improve detection accuracy. From the reviewed paper, we select several latest proposed schemes to present how each method can be used to recognized properties of Sybil attack in every phase. As not all defense machinery can handle all Sybil attack properties, some have implemented privacy protections, and some can work on mobile networks and fast-changing networks. A practical, energy-efficient, versatile defense mechanism that can cover all Sybil attacks properties is highly recommended [21].

11.4   DISCUSSION, RESULTS AND ANALYSIS

In this study, the power load profile data is not grouped based on certain criteria, but it is grouped generally based on the pattern of power usage. Because the number of clusters to be used is unknown, the authors decided to use several clusters for the K-Means method grouping test. The cluster set used is from 2 cluster sets to 6 cluster sets. After calculating using the K-Means method [22], the optimization level of each cluster is then calculated using the DBI method. Following are the results of calculations on the last iteration as shown in Tables 11.5–11.12.

TABLE 11.5
The Number of Members of Each Iteration Set Cluster 6

Iteration	Number of Members
1	Cluster 1: 5 Member
	Cluster 2: 55 Member
	Cluster 3: 31 Member
	Cluster 4: 2 Member
	Cluster 5: 9 Member
	Cluster 6: 1 Member
2	Cluster 1: 4 Member
	Cluster 2: 51 Member
	Cluster 3: 35 Member
	Cluster 4: 1 Member
	Cluster 5: 10 Member
	Cluster 6: 2 Member

TABLE 11.6
The End Centroid Value of Cluster 1 in Cluster Set 6

A	B	C	D	E	F	G
0.407	0.493	0.548	0.439	0.513	0.491	0.479
H	**I**	**J**	**K**	**L**	**M**	**N**
0.418	0.538	0.433	0.399	0.373	0.267	0.288

TABLE 11.7
The End Centroid Value of Cluster 2 in Cluster Set 6

A	B	C	D	E	F	G
0.062	0.064	0.076	0.104	0.075	0.111	0.074
H	**I**	**J**	**K**	**L**	**M**	**N**
0.108	0.074	0.109	0.081	0.112	0.072	0.068

TABLE 11.8
The End Centroid Value of Cluster 3 in Cluster Set 6

A	B	C	D	E	F	G
0.155	0.148	0.188	0.197	0.19	0.203	0.188
H	**I**	**J**	**K**	**L**	**M**	**N**
0.202	0.191	0.202	0.191	0.206	0.167	0.156

TABLE 11.9
The End Centroid Value of Cluster 4 in Cluster Set 6

A	B	C	D	E	F	G
0.641	0.788	1.608	1.329	1.635	1.558	1.690
H	**I**	**J**	**K**	**L**	**M**	**N**
1.633	1.653	1.671	1.729	1.640	0.719	1.351

TABLE 11.10
The End Centroid Value of Cluster 5 in Cluster Set 6

A	B	C	D	E	F	G
0.333	0.274	0.348	0.277	0.35	0.279	0.337
H	**I**	**J**	**K**	**L**	**M**	**N**
0.277	0.348	0.276	0.332	0.283	0.328	0.293

TABLE 11.11

The End Centroid Value of Cluster 6 in Cluster Set 6

A	B	C	D	E	F	G
0.359	0.480	1.043	1.125	1.045	1.219	1.020

H	I	J	K	L	M	N
1.200	1.050	1.200	1.034	1.269	0.434	0.646

TABLE 11.12

The Results of the DBI Value Calculation for Each Set of Clusters

Set Cluster	DBI	Number of Members
2	1.234	Cluster 1: 15
		Cluster 2: 88
3	0.931	Cluster 1: 4
		Cluster 2: 58
		Cluster 3: 41
4	0.893	Cluster 1: 12
		Cluster 2: 54
		Cluster 3: 34
		Cluster 4: 3
5	1.174	Cluster 1: 4
		Cluster 2: 51
		Cluster 3: 35
		Cluster 4: 3
		Cluster 5: 40
6	0.990	Cluster 1: 4
		Cluster 2: 51
		Cluster 3: 35
		Cluster 4: 1
		Cluster 5: 10
		Cluster 6: 2

Based on the calculation results in the application, the cluster set is the most optimal is set cluster 4 because it has the smallest DBI value, that is, 0.893, that means set cluster 4 has the density of each object with the best centroid and the distance between the clusters is also well separated [23].

After getting the optimal set of clusters, next is the testing phase. At this step, the data being tested is data of 3 customers categorized as customers with non-normal usage of electricity power. The test is, by determining the distance of each data testing object to each centroid in the cluster 4 set then the 3 data are tested in the application with the output that is, the 3rd data is not normal in electricity power usage. The test is, by determining the distance of each data testing object to each centroid in the cluster 4 set then the 3 data are tested in the application with the output that is, all 3 data are classified into customers with unnatural usage because based on the data allocation process to the centroid set the closest cluster 4, the distance of the testing data exceeds the maximum distance of each object in each cluster in the cluster 3 set [24,25].

11.4.1 GENERAL DETECTION ISSUE

As a general need for defense machinery in the wireless ad-hoc network to be integrated with IoT, several issues arise both related to the accuracy of defense machinery, the possibility of implementation, and others. Some issues related to this include [26]:

- Accuracy: defense mechanism can detect Sybil at each phase with different properties. It must be able to discover large percentage of Sybil nodes to eliminate damage.
- Cooperative Sybil detection: to detect effectively, all nodes in networks participate independently in the Sybil node detection process [27].
- Low overhead costs: the proposed approach works more efficiently and requires fewer system resources.
  - Does not need for additional hardware at high prices.
  - Does not increase message exchange on the network.
  - Does not require much memory.
- Detection time: the time needed to find and delete a Sybil entity is an essential factor that must be minimal.
- Implementation: every IoT implementation such as in industry, smart home, smart grid, and others, there are special needs that must be considered in applying defense machinery [28].

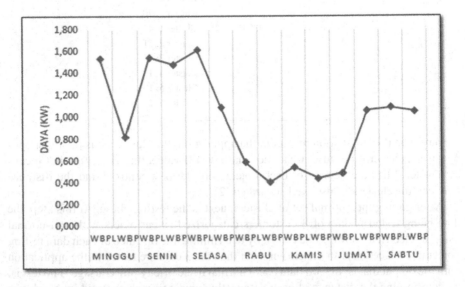

FIGURE 11.3 Graphs of power consumption customers are not normal.

11.4.2 VANET Issue

In the wireless ad-hoc network area, VANET has become the most talked about topic lately, with specific needs that VANETs require additional requirements for security guarantees. Issues discussed in several papers reviewed are [29]:

- Privacy Issue: most vehicle users hope that their identity information can be stored in VANET because they are afraid that their trip will leak with that identity.
- Safety Issue: VANET does not allow a decrease in reputation after a severe traffic accident to prevent another attack, because damage to life and things in this attack cannot be repaired.

11.4.3 LEARNING-BASED ISSUE

Defense machinery in the IoT infrastructure must be prepared with the needs of a "smart" system so that the application of scientific fields on artificial intelligence, especially machine learning is very open. Machine learning that needs to be applied to Sybil's defense machinery include: [30]

- Deep Learning: with the development and the number of entities in an IoT infrastructure, a mechanism based on thorough analysis is needed, and deep learning has been successfully used in various areas including intrusion detection systems.
- Online Learning: most of the data sent on IoT infrastructure, including WANET-based IoT, is a data stream, so online learning needs to be a concern for solutions on detection that continuously enhance the capability of defense machinery.

11.4.4 CENTRALIZED VS. DECENTRALIZED ISSUE

- Centralized issue: some defense machinery uses centralized detection, which requires a trusted center. Several papers on VANET build trust relationships that are bestowed on RSU. Installation of such infrastructure nationally is challenging to achieve in the early stages of VANET. Even in the medium term, there may still be many places that are not covered by RSU [31].
- Decentralized issue: on the mechanism that relies on each node as a detector, all must know the credibility of each node that shares information around it and ensure all messages received are trusted and correct. However, this mechanism can work well assuming that most nodes are trusted nodes [32,33].

11.5 CONCLUSION

Based on the description of the discussion above, the following conclusions can be drawn. Clustering of historical data on customer power usage of AMR business class has been successfully built and can be used to classify and determine the pattern of electric power usage based on the number of cluster sets that have been calculated for their optimization. Based on the results of testing, the application with 2–6 sets of clusters is the most optimal because it has the smallest DBI value (minimum), 0.893. In a set of 4 clusters, cluster 1 has 12 members, cluster 2: 54 members, cluster 3: 34 members, and cluster 4:3 members. There are 3 patterns of electric power usage of AMR customers of business class (B-3 or medium voltage class, power limit above 200 kVA). The design of clustering applications for AMR customers' historical electrical power usage begins by analyzing the current system and drafting a proposed system using a flowchart. In this paper, we have provided a comprehensive review of defense machinery against Sybil attacks, including defining the Sybil attack properties, building the taxonomy of these machinery, and analyzing the problems that still exist in defense machinery against Sybil on wireless ad-hoc networks related to their implementation in IoT. Several challenges have been mentioned to be implemented in a practical IoT system.

CONFLICT OF INTEREST

On behalf of all authors, the corresponding author states that there is no conflict of interest.

REFERENCES

1. Bhatti SS, Umair EM, Lodhi U, Haq S. Electric-power-transmission-and-distribution-losses-overview-and-minimization-in-Pakistan.docx. *International Journal of Scientific and Engineering Research*. 2015;6(4)1–8.
2. Pradana B, Purba P, Warman E. Pendekatan Kurva Beban Pada Jaringan Distribusi PT. PLN (Persero) Rayon Medan Kota. *Singuda Ensikom*. 2014;6(2):60–4.
3. Reina DG, Toral SL, Barrero F, Bessis N, Asimakopoulou E. The Role of Ad Hoc Networks in the Internet of Things: A Case Scenario for Smart Environments. In: Bessis N, Xhafa F, Varvarigou D, Hill R, Li M, editors. *Internet of Things and Inter-cooperative Computational Technologies for Collective Intelligence*. Berlin, Heidelberg: Springer Berlin Heidelberg; 2013 [cited 2019 Aug 17]. pp. 89–113. (Studies in Computational Intelligence).
4. Lu Tan, Neng Wang. Future internet: The Internet of Things. In: *2010 3rd International Conference on Advanced Computer Theory and Engineering (ICACTE)*; 2010. pp. V5–376.
5. Yuntyansyah PA, Wibawa U, Utomo T. Studi Perkiraan Susut Teknis dan Alternatif Perbaikan Pada Penyulang Kayoman Gardu Induk Sukorejo. *Jurnal Mahasiswa Teknik Elektro Universitas Brawijaya*. 2015; 3(1):1–8.
6. Suryanto J. Analisa Perbandingan Pengelompokkan Curah Hujan 15 Harian Provinsi Diy Menggunakan Fuzzy Clustering Dan K-Means Clustering. *Journal of Agroforestry*. 2017;XVI:229–42.
7. Vasudeva A, Sood M. Survey on Sybil attack defense machinery in wireless ad hoc networks. *Journal of Network and Computer Applications*. 2018 Oct 15;120:78–118.
8. Abdulkader ZA, Abdullah A, Abdullah MT, Zukarnain ZA. A survey on Sybil attack detection in vehicular ad hoc networks (VANET). *Journal of Computers (Taiwan)*. 2018;29(2):1–6.
9. Newsome J, Shi E, Song D, Perrig A. The Sybil attack in sensor networks: Analysis & defenses. In: *Proceedings of the 3rd International Symposium on Information Processing in Sensor Networks*. New York, NY, USA: ACM; 2004. pp. 259–68. (IPSN '04).

10. Kitchenham B, Pearl Brereton O, Budgen D, Turner M, Bailey J, Linkman S. Systematic literature reviews in software engineering - A systematic literature review. *Information and Software Technology*. 2009 Jan;51[1]:7–15.
11. Ghosh S, Kumar S. Comparative analysis of K-means and fuzzy C-means algorithms. *International Journal of Advanced Computer Science and Applications*. 2013; 4(4):35–39.
12. Rahman AT., Wiranto, Rini A. Coal trade data clustering using K-means (case study Pt. Global Bangkit Utama). *ITSMART Jurnal Teknologi dan Informasi*. 2017; 6(1):1–8.
13. Mishra AK, Tripathy AK, Puthal D, Yang LT. Analytical model for Sybil attack phases in Internet of Things. *IEEE Internet of Things Journal*. 2019;6(1, SI):379–87.
14. Feng X, Li C, Chen D, Tang J. A method for defensing against multi-source Sybil attacks in VANET. *Peer-to-Peer Networking and Applications*. 2017 Mar;10(2):305–14.
15. Amuthavalli R, Bhuvaneswaran RS. Detection and prevention of Sybil attack in wireless sensor network employing random password comparison method. *Journal of Theoretical and Applied Information Technology*. 2014;67(1):236–46.
16. Sharma P, Gupta G. Proficient techniques and protocols for the identification of attacks in WSN: A review. *Indian Journal of Science and Technology*. 2016;9(42):1–4.
17. Nirmal Raja K, Maraline Beno M. Secure data aggregation in wireless sensor network-Fujisaki Okamoto (FO) authentication scheme against Sybil attack. *Journal of Medical Systems*. 2017;41(7): 1–6.
18. Saud Khan M, Khan NM. Low complexity signed response based Sybil attack detection mechanism in wireless sensor networks. *Journal of Sensors*. 2016;2016:1–9.
19. Vinayagam SS, Parthasarathy V. IPTTA: Leveraging token-based node IP assignment and verification for WSN. In: *2014 International Conference on Science Engineering and Management Research (ICSEMR)*. 2014.
20. Shu J, Liu X, Yang K, Zhang Y, Jia X, Deng RH. SybSub: Privacy-preserving expressive task subscription with Sybil detection in crowdsourcing. *IEEE Internet of Things Journal*. 2019 Apr;6(2):3003–13.
21. de Sales TBM, Perkusich A, de Sales LM, de Almeida HO, Soares G, de Sales M. ASAP-V: A privacy-preserving authentication and Sybil detection protocol for VANETs. *Information Sciences*. 2016;372:208–24.
22. Kuswantoro E, Suprapto YK. Penerapan Algoritma k-Means Dengan Optimasi Jumlah Cluster Untuk Pengelompokan Angkatan Kerja Propinsi Jatim. *JAVA Journal of Electrical and Electronics Engineering*. 2015; 13(1):1–5.
23. Muhammad A, Tumaliang H, Silimang S. Analisa Rugi-Rugi Energi Listrik Pada Jaringan Distribusi (JTM) Di PT. PLN (Persero) Area Gorontalo. *Anal. Rugi-Rugi Energi List. Pada Jar. Distrib. Di PT. PLN Area Gorontalo*. 2019; 13(2):157–167.
24. Nasution FE, T JM. Pengaruh Faktor Daya Customer Industri Terhadap Rugi – Rugi Pada Jaringan Sisi Sekunder Transformator Distribusi PT. PLN (Persero) Area Serpong. *Semin. Nas. Teknol. Fak. Tek. Univ. Krisnadwipayana*. 2019; 148–162.
25. Heriyanto A. Studi Kasus Kinerja AMR (Automatic Meter Reading) Pada Customer Potensial Daya 41.5 KVA – 200 KVA Di Situbondo. Jurnal Teknik Elektro Universitas Muhammadiyah Jember Tahun 2016, Page No.1-13 2016.
26. Feng X, Tang J. Obfuscated RSUs vector based signature scheme for detecting conspiracy Sybil attack in VANETs. *Mobile Information Systems*. 2017;2017: 1–11.
27. Soni M, Jain A. Secure communication and implementation technique for Sybil attack in vehicular ad-hoc networks. In: *Proceedings of the 2nd International Conference on Computing Methodologies and Communication, ICCMC 2018*; 2018. pp. 539–43.
28. Reddy DS, Bapuji V, Govardhan A, Sarma SSVN. Sybil attack detection technique using session key certificate in vehicular ad hoc networks. In: *2017 International Conference on Algorithms, Methodology, Models and Applications in Emerging Technologies (ICAMMAET)*; 2017. pp. 1–5.

29. Sharma AK, Saroj SK, Chauhan SK, Saini SK. Sybil attack prevention and detection in vehicular ad hoc network. In: Astya, PN and Swaroop, A and Sharma, V and Singh, M, editors. *2016 IEEE International Conference on Computing, Communication and Automation (ICCCA).* IEEE; IEEE UP Sect; IEEE Uttar Pradesh Sect SP C Chapter; Galgotias Univ, Sch Comp Sci & Engn; 2016. pp. 594–9.

30. Hussain R, Oh H. On secure and privacy-aware Sybil attack detection in vehicular communications. *Wireless Personal Communications.* 2014 Aug;77(4):2649–73.

31. Alimohammadi M, Pouyan AA. Sybil attack detection using a low-cost short group signature in VANET. In: *2015 12TH International Iranian Society of Cryptology Conference on Information Security and Cryptology (ISCISC).* Iranian Soc Cryptol; 2015. pp. 23–8.

32. Mangundap J, Silimang, S, Tumaliang H. Analisa Rugi-Rugi Daya Jaringan Distribusi Di PT. PLN (Persero) Area Manado 2017. *J. Tek. Elektro dan Komput.* 2018; 7(3):219–226.

33. Agustina E, Amalia AF. Penurunan Susut Non Teknis Pada Jaringan Distribusi Menggunakan Sistem Automatic Meter Reading DI PT. PLN (Persero). *J. Tek. Mesin.* 2017; 5:37–39.

Index

Printed in the United States
by Baker & Taylor Publisher Services

Printed in the United States
by Baker & Taylor Publisher Services